Neelam Dewangan

A detailed Study of 4G in Wireless Communication

Looking insight in issues in OFDM

Anchor Academic Publishing

Dewangan, Neelam: A detailed Study of 4G in Wireless Communication: looking insight in issues in OFDM. Hamburg, Anchor Academic Publishing 2013

Buch-ISBN: 978-3-95489-084-2
PDF-eBook-ISBN: 978-3-95489-584-7
Druck/Herstellung: Anchor Academic Publishing, Hamburg, 2013

Bibliografische Information der Deutschen Nationalbibliothek:
Die Deutsche Nationalbibliothek verzeichnet diese Publikation in der Deutschen Nationalbibliografie; detaillierte bibliografische Daten sind im Internet über http://dnb.d-nb.de abrufbar.

Bibliographical Information of the German National Library:
The German National Library lists this publication in the German National Bibliography. Detailed bibliographic data can be found at: http://dnb.d-nb.de

All rights reserved. This publication may not be reproduced, stored in a retrieval system or transmitted, in any form or by any means, electronic, mechanical, photocopying, recording or otherwise, without the prior permission of the publishers.

Das Werk einschließlich aller seiner Teile ist urheberrechtlich geschützt. Jede Verwertung außerhalb der Grenzen des Urheberrechtsgesetzes ist ohne Zustimmung des Verlages unzulässig und strafbar. Dies gilt insbesondere für Vervielfältigungen, Übersetzungen, Mikroverfilmungen und die Einspeicherung und Bearbeitung in elektronischen Systemen.

Die Wiedergabe von Gebrauchsnamen, Handelsnamen, Warenbezeichnungen usw. in diesem Werk berechtigt auch ohne besondere Kennzeichnung nicht zu der Annahme, dass solche Namen im Sinne der Warenzeichen- und Markenschutz-Gesetzgebung als frei zu betrachten wären und daher von jedermann benutzt werden dürften.

Die Informationen in diesem Werk wurden mit Sorgfalt erarbeitet. Dennoch können Fehler nicht vollständig ausgeschlossen werden und die Diplomica Verlag GmbH, die Autoren oder Übersetzer übernehmen keine juristische Verantwortung oder irgendeine Haftung für evtl. verbliebene fehlerhafte Angaben und deren Folgen.

Alle Rechte vorbehalten

© Anchor Academic Publishing, Imprint der Diplomica Verlag GmbH
Hermannstal 119k, 22119 Hamburg
http://www.diplomica-verlag.de, Hamburg 2013
Printed in Germany

Table of contents

CHAPTER 1 1

Introduction 1

   1.1 Introduction to Long term evolution (LTE) 1

      1.2   Technologies involved 3

        1.2.   OFDM (Orthogonal Frequency Division Multiplexing) 3

        1.2.3.   OFDMA (Orthogonal Frequency Division Multiple Access) 3

        1.2.3   MIMO (Multiple Input Multiple Output) 4

      1.2.4 SC-FDMA (Single Carrier Frequency Division Multiple Access) 4

      1.3   Brief History of OFDM 4

        1.3.1 Multipath Channels 5

   1.4 Basic Concepts 6

      1.4.*1 Frequency* Division Multiplexing (FDM) 6

      1.4.2 Time Division Multiplexing (TDM) 6

      1.4.3 Orthogonal Frequency Division Multiplexing (OFDM) 7

   1.5 Introduction to OFDM 7

      1.5.1 Orthogonal Frequency Division Multiplexing (OFDM) 7

      1.5.2 OFDM is a special case of FDM 10

   1.6 SC-FDMA and OFDMA Tx-Rx Structure 11

   1.7 Inter - Symbol Interference(ISI) 13

   1.8 Inter - Carrier Interference 13

   1.9 Understanding Concept of Cyclic Prefix 13

   1.10 OFDM using Inverse DFT 16

   1.11 Advantages of OFDM 17

   1.12 Disadvantages of OFDM 17

   1.13 Peak to Average Power Ratio 18

   1.14 PAPR Reduction Techniques 18

CHAPTER 2 20

Literature Review 20

   2.1 Different methods for Peak-to-Average Power (PAPR) Reduction in Orthogonal Frequency Division Multiplexing (OFDM) 20

CHAPTER 3 28

PROBLEM IDENTIFICATION 28

3.1 Clipping and Filtering.. 31

3.2 Coding... 32

3.3 Interleaving... 33

3.4 Companding 34

3.5 Peak Windowing... 34

3.6 Additive Correcting Function.. 34

3.7 Selected Mapping (SLM) 35

3.8 Tone Reservation.. 35

3.9 Tone Injection... 35

3.10 Selective Scrambling (Interleaving) 36

CHAPTER 4 .... **37**

METHODOLOGY **37**

4.1 Objectives 37

4.2 Hardware and Software Required 37

    4.2.1 Hardware Required 37

    4.2.2 Software Required 37

4.3 Simulation model of OFDM System 38

    4.3.1 Random Data Generator 38

    4.3.2 Serial to Parallel Conversion... 39

    4.3.3 Modulation of Data 39

    4.3.4 Inverse Fourier Transform 39

    4.3.5 Guard Period 39

    4.3.6 Parallel to Serial Converter... 39

    4.3.7 Channel... 39

    4.3.8 Receiver 40

4.4 Calculation of PAPR and CCDF of Original OFDM Signal 40

4.5 Complimentary Cumulative Distribution Function (CCDF)....................................... 41

4.6 Calculation of SNR and BER of Original OFDM Signal... 42

    4.6.1 Additive White Gaussian Noise(AWGN) Channel... 42

    4.6.2 Signal-to-Noise Ratio (SNR) 43

    4.6.3 Bit Error Rate (BER) 44

4.7 Criteria for selection of PAPR reduction techniques.. 44

4.8 Definition of Efficient PAPR 46

CHAPTER 5	47
PAPR REDUCTION TECHNIQUES	47
5.1 SELECTIVE MAPPING	48
5.2 Clipping - Based Active Constellation Extension Algorithm	49
5.2.1 Limitations of CB-ACE Algorithm	52
5.3 Exponential Companding Transform	52
5.3.1 Companding of Original OFDM Signal by using Exponential Companding Transform	53
5.3.2 Advantages of Exponential Companding Transform	54
5.3.3 Limitations of Exponential Companding Transform	54
5.4 Adaptive Active Constellation Extension Algorithm	54
CHAPTER 6	58
PROPOSED METHOD	58
6.1 Selected Mapping With Riemann Matrix	58
6.2 Concept of Riemann matrices	59
CHAPTER 7	63
RESULTS AND DISCUSSION	63
7.1 PAPR vs CCDF of Original OFDM Signal	63
7.2 BER of Original OFDM Signal	64
7.3 PAPR vs CCDF of OFDM Signal by using Selective Mapping (SLM) Technique	65
7.4 CCDF Plot for Clipping-Based Active Constellation Extension (CB-ACE) Technique	67
7.5 BER Plot for Clipping-Based Active Constellation Extension (CB-ACE) Technique	68
7.6 CCDF Plot for Adaptive Active Constellation Extension (Adaptive -ACE) Technique	70
7.7 BER Plot for Adaptive Active Constellation Extension (Adaptive -ACE) Technique	71
7.8 CCDF Plot for Exponential Companding Technique	72
7.9 BER Plot for Exponential Companding Technique	73
7.10 CCDF Plot for Proposed Technique- SLM with Riemann Matrix	74
7.11 BER Plot for Proposed Technique- SLM with Riemann Matrix	75
CHAPTER 8	77
CONCLUSION AND FUTURE SCOPE	77
REFERENCES	83

CHAPTER 1

Introduction

1.1 Introduction to Long term evolution (LTE)

The ever increasing demand for high data rates in wireless communications systems has arisen in order to support broadband services. Long term evolution (LTE) is standardized by the third generation partnership project (3GPP) and is an evolution of existing 3G technologies in order to meet projected customer needs over the next decades. Current working assumptions in 3GPP LTE are to use orthogonal frequency division multiplexing access (OFDMA) for downlink and single carrier-frequency division multiple access (SC-FDMA) for uplink. SC-FDMA is a promising technique for high data rate transmission that utilizes single carrier modulation and frequency domain equalization. Single carrier transmitter structure leads to keep the peak-to-average power ratio (PAPR) as low as possible that will reduce the energy consumption. SC-FDMA has similar throughput performance and essentially the same overall complexity as OFDMA. [29]

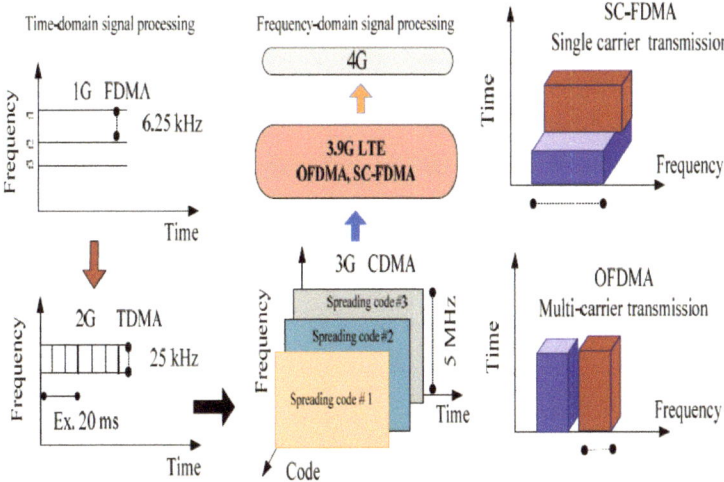

Figure 1.1. Multiple access schemes [28]

Recently, OFDMA has received much attention due to its applicability to high speed wireless multiple access communication systems. The problems associated with OFDM, however, are also inherited by OFDMA. Hence, OFDMA also suffers from high PAPR. In OFDMA process the whole data block is treated as one unit. OFDMA systems are more difficult since only part of the subcarriers in one OFDMA data block are of demodulated by each user's receiver. If downlink PAPR reduction is achieved by schemes designed for OFDM, each user has to process the whole data block, then each user demodulates the assigned sub-carriers meant for them and extract their own information. This introduces additional processing for each user's receiver. Various statistical PAPR characteristics and PAPR reduction in OFDM signals are analyzed in different research papers and various approaches have been proposed to reduce the PAPR including amplitude clipping, selected mapping technique, coding schemes, tone reservation technique etc. [29]

3GPP is a standardization committee that has produced several specification documents for LTE. The different targets of LTE are [30]:

- **Peak Data Rates:** Evolved Universal Terrestrial Radio Access (E-UTRA) is expected to support significantly increased instantaneous peak data rates. The peak data rates may depend on the number of transmit and receive antennas at the User Equipment (UE). For this baseline configuration, the system should support an instantaneous downlink peak data rate of 100 Mb/s within a 20 MHz downlink spectrum allocation and an instantaneous uplink peak data rate of 50 Mb/s within a 20 MHz uplink spectrum allocation.

- **Latency:** It is expected that at each user plane, latency should be less than 5 ms one-way and a control plane transition time of less than 50 ms from dormant to active mode and less than 100 ms from idle to active mode.

- **User throughput:** It is expected that 2-3 times higher downlink throughput than HSDPA.

- **Spectrum efficiency:** 3-4 times higher spectrum efficiency (in bits/s/Hz/site) in downlink and 2-3 times higher in uplink, compared to Release 6 High-Speed Downlink Packet Access (HSDPA).

- **Mobility:** LTE should support mobility across the cellular network and should be optimized for 0 to 15 km/h. Furthermore, should also support higher performance at 15 and 120 km/h. Connection shall be maintained at speeds from 120 km/h to 350 km/h (or even up to 500 km/h depending on the frequency band).

- **Coverage:** Cell ranges up to 5 km support the above targets; up to 30 km will suffer some degradation in throughput and spectrum efficiency and up to 100 km will have overall performance degradation.

- **Spectrum flexibility:** LTE should support several different spectrum allocation sizes as: 1.25 MHz, 1.6 MHz, 2.5 MHz, 5 MHz, 10 MHz, 15 MHz and 20 MHz with both TDD and FDD modes. Shall also enable the flexibility to modify the radio resource allocation for broadcast transmission according to specific demand or operator's policy. [30]

1.2 Technologies involved

LTE employs different technologies such as OFDM, OFDMA, MIMO and SC-FDMA. These methods are briefly described in the following subsections [32].

1.2. OFDM (Orthogonal Frequency Division Multiplexing)

OFDM is a digital multi-carrier modulation scheme that distributes the data over a large number of carriers closely spaced. The two main characteristics are that each subcarrier is modulated using varying levels of QAM modulation and each OFDM symbol is preceded by a cyclic prefix (CP) used to effectively eliminate intersymbol interference (ISI).

OFDM has several advantages such as can easily adapt to severe channel conditions, is robust against ISI and fading caused by multipath and provide high spectral efficiency. But it also has disadvantages as it is sensitive to Doppler shift, defined as the change in frequency of a wave for an observer moving relative to the source of the waves. It is also sensitive to frequency synchronization problems and having high peak-to-average-power ratio (PAPR).

1.2.3. OFDMA (Orthogonal Frequency Division Multiple Access)

Orthogonal Frequency Division Multiple Access (OFDMA) is a multi-user version of OFDM. Multiple access is achieved by assigning different OFDM sub-channels to different users. Among the advantages of OFDMA, one important property is having its robustness to fading and interference. It also averages the interferences within the cells using allocation with cyclic permutation and offers frequency diversity by spreading the carriers all over the used spectrum. On the other hand, OFDMA is higher sensible to frequency offsets and phase noise .Moreover, the resistance to the frequency-selective fading may partly be lost if very few subcarriers are assigned to each user and if the same carrier is used in every OFDM symbol. OFDMA is used as the multiplexing scheme in the LTE downlink.

1.2.3 MIMO (Multiple Input Multiple Output)

MIMO technology offers significant increases in data throughput and link range without additional bandwidth or transmitted power. There are multiple transceivers at both the base station and UE in order to enhance link robustness and increase data rates for the LTE downlink.

1.2.4 SC-FDMA (Single Carrier Frequency Division Multiple Access)

LTE requirements in uplink differ in several aspects from downlink. The main fact is the transmission scheme used. Power consumption is a key consideration for UE terminals and for this; the high PAPR and related loss of efficiency associated to OFDM signaling are major concerns. As a result, an alternative to OFDM was sought for use in the LTE uplink.

The solution is Single Carrier – Frequency Domain Multiple Access (SC-FDMA) that suits very well with the LTE uplink requirements. The basic transmitter and receiver architecture is very similar (nearly identical) to OFDMA as shown in Figure 2., and it offers the same degree of multipath protection [32].

1.3 Brief History of OFDM

The idea of Orthogonal Frequency Division Multiplexing (OFDM) was proposed in mid 1960's which used parallel data transmission and Frequency Division Multiplexing. In the 1960's the OFDM was used in several high frequency military systems .In 1971 Weinstein and Ebert applied the Discrete Fourier Transform to parallel data transmission systems as a part of modulation and Demodulation process. In 1980's OFDM was studied for high speed modem digital mobile communication and high density recording in which pilot tone was used to stabilize carrier and Frequency control and Trellis code was implemented which gave rise to Coded-OFDM.In 1780, Hirosaki suggested an equalization algorithm in order to suppress both intersymbol and intercarrier interference caused by the channel impulse response or timing and frequency errors. In 1980 , Hirosaki introduced the DFT –based implementation of Saltzburg's O-QAM OFDM system. In 1990s, OFDM has been used extensively for wideband data communication over mobile radio FM channels, high-bit-rate digital subscriber lines (HDSL, 1.6 Mb/s), asymmetric digital subscriber lines (ADSL, 1536 Mb/s), very high-speed digital subscriber lines (VHDSL, 100 Mb/s), digital audio broadcasting (DAB) and HDTV terrestrial broadcasting. OFDM is used in wireless digital radio, TV transmissions, particularly in Europe, also used in wireless Local Area Networks (LANs) as specified by the IEEE 802.11, IEEE 802.16, IEEE802.20 and the European Telecommunications Standards Institute (ETSI) HiperLAN/2 standards.[27]

1.3.1 Multipath Channels

The transmitted signal faces various obstacles and surfaces of reflection, as a result of which the received signals from the same source reach at different times. This gives rise to the formation of "echoes" which affect the other incoming signals. Dielectric constants, permeability, conductivity and thickness are the main factors affecting the system. Multipath channel propagation is devised in such a manner that there will be a minimized effect of the echoes in the system in an indoor environment. Measures are needed to be taken in order to minimize echo in order to avoid ISI.

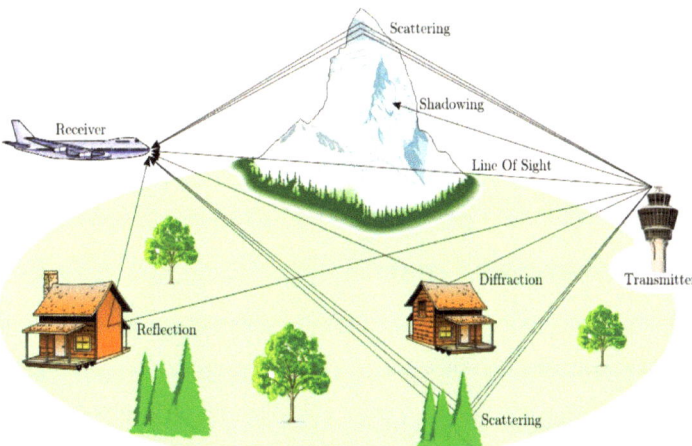

Figure 1.3.1 Multipath Propagation

1.4 Basic Concepts

1.4.1 Frequency *Division Multiplexing* (FDM)

Frequency Division Multiplexing is being used for a long time to carry the data more than one carrier signal. It divides the total channel bandwidth into sub channels so that each sub channel carries the modulated data into a separate carrier frequency. There will be some guard bands between the adjacent channels so that there is no inter channel interference .FDM technique are quite popular technique used in telephones line.

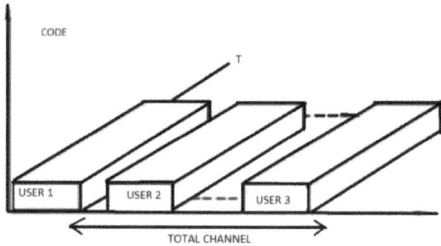

Figure 1.4.1 Frequency Division Multiplexing

1.4.2 Time Division Multiplexing (TDM)

Time Division Multiplexing is another efficient technique which improves the capacity by splitting the frequencies in different time slots. It allows the user to access the entire frequency band at a particular instant of time. Other users share the same frequency channel at different time slot. TDMA system divide the radio spectrum into time slots, and in each slot only one user is allowed to transmit and receive

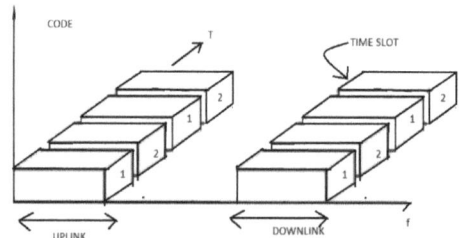

Figure 1.4.2 Time Division Multiplexing

1.4.3 Orthogonal Frequency Division Multiplexing (OFDM)

In order to solve the bandwidth efficiency problem, Orthogonal Frequency Division Multiplexing (OFDM) technique is used. OFDM is a multicarrier transmission technique, which divides the total available bandwidth into many subcarriers; each subcarrier is modulated by a low rate data stream. In term of multiple access technique OFDM is similar to FDMA such that the multiple users access is achieved by subdividing the available bandwidth into multiple channels. However, OFDM uses the spectrum much more efficiently by spacing the channels much closer together. This is achieved by making all the carriers orthogonal to one another, preventing interference between the closely spaced carriers. The orthogonality of the carriers means that each carrier has an integer number of cycles over a symbol period. Due to this, the spectrum of each carrier has a null at the center frequency of each of the other carriers in the system. This results in no interference between the carriers, allowing to be spaced as close as theoretically possible. This overcomes the problem of overhead carrier spacing required in FDMA. [27]

1.5 Introduction to OFDM

1.5.1 Orthogonal Frequency Division Multiplexing (OFDM)

Orthogonal Frequency Division Multiplexing (OFDM) is a method of Digital Modulation in which a signal is split into several narrowband channels at different frequencies. The OFDM technology was first conceived in the 1960s and 1970s during the research into minimizing interference among the channels near each other in frequency. The main idea behind the OFDM is that since low-rate modulations are less sensitive to multipath, the better way is to send a number of low rate streams in parallel than sending one high rate waveform. This can be exactly done in OFDM. The OFDM divides the frequency spectrum into sub-bands small enough so that the channel effects are constant (flat) over a given sub-band. Then a classical IQ modulation (BPSK, QPSK, M-QAM, etc.) is sent over the sub-band. If designed correctly, all the fast changing effects of the channel disappear as they are now occurring during the transmission of a single symbol and are thus treated as flat fading at the received.

A large number of closely spaced orthogonal subcarriers are used to carry data. The data is divided into several parallel data streams or channels, one for each subcarrier.Each subcarrier is modulated with a conventional modulation scheme such as Quadrature Amplitude Modulation (QAM) or Phase Shift Keying (PSK) at a low symbol rate. The total data rate is to be maintained similar to that of the conventional single carrier modulation scheme with the same bandwidth. Orthogonal Frequency Division Multiplexing (OFDM) is a promising technique for achieving high data rates and combating multipath fading in Wireless Communications.

The independent sub-channels can be multiplexed by frequency division multiplexing called as multi carrier transmission and if they are multiplexed by Code division multiplexing then it is called multi-code transmission.

There is a precise mathematical relationship between the frequencies of the carriers. It is possible to arrange the carriers in an OFDM signal so that the sidebands of the individual carriers overlap and the signals can still be received without adjacent carrier interference. In order to do this the carriers must be mathematically orthogonal. The carriers are linearly independent (i.e. orthogonal) if the carrier spacing is a multiple of $1/T_s$ where T_s is the symbol duration. Figure 1.5.1(b) shows the minimum frequency difference required for carriers to be orthogonal. The OFDM system transmits a large number of narrowband carriers, which are closely spaced. [47]

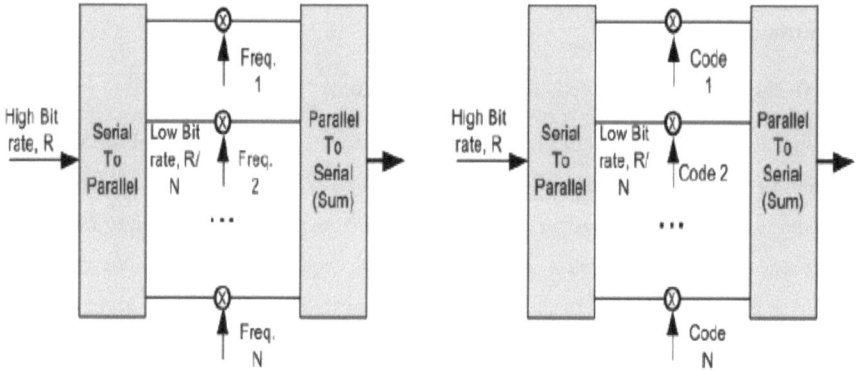

Figure 1.5.1 (a) Multi-carrier FDM and Multi-Code Division Multiplex

If a sine wave of frequency a multiplied by a sinusoid (sine or cosine) of a frequency b,

$$f(t) = sin2\pi at \times sin2\pi bt \qquad (1)$$

Here both a and b are integers, since these two components are each a sinusoid, the integral is equal to zero over one period. The integral or area under this product is given by

$$= \int_0^{2\pi} \sin 2\pi at \times \sin 2\pi bt$$

$$= \frac{1}{2}\int_0^{2\pi} \cos(a-b)wt - \frac{1}{2}\int_0^{2\pi} \cos(a+b)wt = 0 \quad (2)$$

Since the carriers are all sine/cosine wave, we know that area under one period of a sine or a cosine wave is zero. So when a sinusoid of frequency n multiplied by a sinusoid of frequency m/n, the area under the product is zero. In general for all integers n and m, sin mx, cos mx, cos nx, sin nx are all orthogonal to each other. These frequencies are called harmonics.

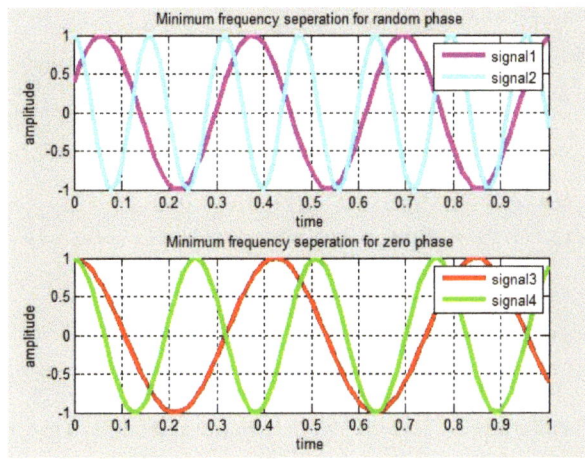

Figure 1.5.1 (b) minimum frequency difference required for carriers to be orthogonal

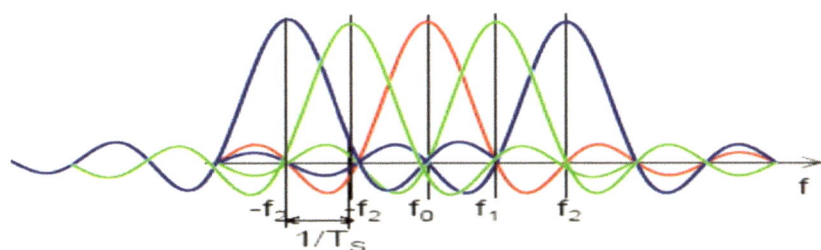

Figure 1.5.1 (c) Example of OFDM spectrum for 5 orthogonal carriers

The orthogonality allows simultaneous transmission of many sub-carriers in a tight frequency space without interference from each other. This is similar to CDMA, where codes are used to make data sequences independent (also orthogonal) which allows many independent users to transmit in same space successfully.[13]

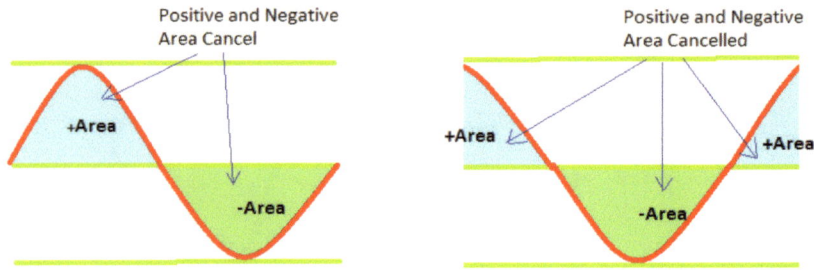

Figure 1.5.1 (d) The area under a sine and a cosine wave over one period is always zero.

1.5.2 OFDM is a special case of FDM

Orthogonal Frequency Division Multiplexing (OFDM) is a special case of Frequency Division Multiplexing(FDM). Suppose a bandwidth goes from frequency say a to b, then this can be subdivided into a frequency of equal spaces. In frequency space the modulated carriers would look like this.

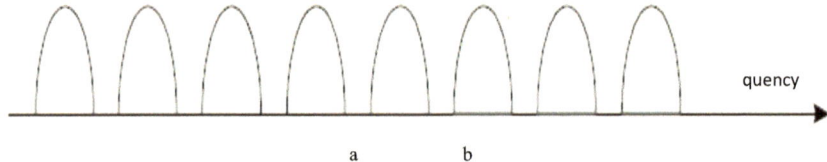

Figure 1.5.2 (a) Bandwidth utilization in Frequency Division Multiplexing

The frequencies a and bcan be anything, integer or non-integer since no relationship is implied between a and bsame is true of the carrier center frequencies which are based on frequencies that do not have any special relationship to each other. If frequency C_1 and C_n were such that for any n, an integer, the following holds.

$$C_n = nC_1$$

So that,
$$C_2 = 2C_1$$
$$C_3 = 3C_1$$
$$C_4 = 4C_1$$

All three of these frequencies are harmonic to C_1.

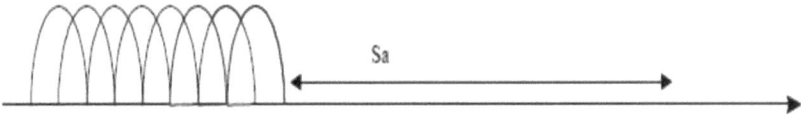

Figure 1.5.2 (b) Bandwidth utilization of OFDM

In this case, these carriers are orthogonal to each other. Hence when added together, they do not provide interference with each other. In FDM, frequenciesare not orthogonal to each other, so it suffers interference from neighbor carriers. To provide adjacent channel interference protection, signals are moved further apart. Each carrier is separated apart by a 10% guard band. It is the guard band that keeps interference under control. [27]

1.6 SC-FDMA and OFDMA Tx-Rx Structure

The transmitter of an SC-FDMA system converts a binary input signal to a sequence of modulated subcarriers. To do so, it performs the signal processing operations as shown in Figure 1.6. Signal processing is repetitive in a few different time intervals. Resource assignment takes place in transmit time intervals (TTIs). The TTI is further divided into time intervals referred to as blocks. A block is the time used to transmit all of subcarriers once. At the input to the transmitter, a baseband modulator transforms the binary input to a multilevel sequence of complex numbers xn in one of several possible modulation formats including binary phase shift keying (BPSK), quaternary PSK (QPSK), 16 level quadrature amplitude modulation (16-QAM) and 64-QAM.

SC - FDMA

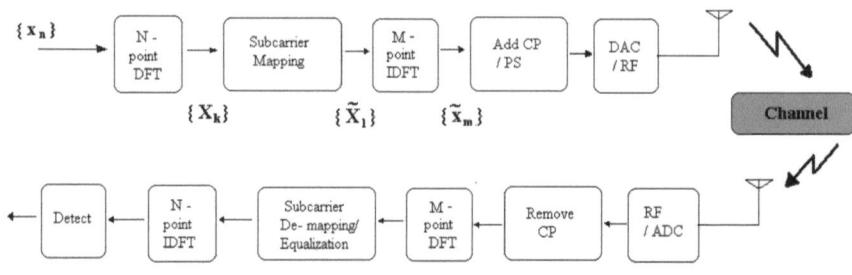

OFDMA

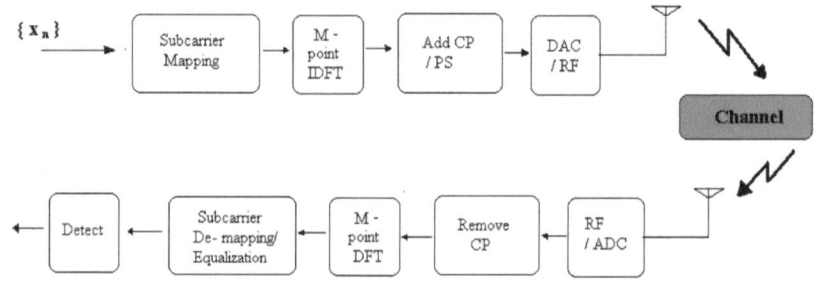

*CP : Cyclic Prefix , PS : Pulse Shaping

Figure 1.6: SC-FDMA and OFDMA Tx-Rx Structure [18]

Next, the transmitter groups the modulation symbols, x_n into blocks, each containing N symbols. The first step in modulating the SC-FDMA subcarriers is to perform an N-point Discrete Fourier Transform (DFT), to produce a frequency domain representation X_k of the input symbols. Then mapping of each of the N DFT outputs to one of the M (> N) orthogonal subcarriers that can be transmitted. As in OFDMA, a typical value of M is 256 subcarriers and $N = M/Q$ is an integer submultiple of M. Q is the bandwidth expansion factor of the symbol sequence.

The transmitter performs two other signal processing operations prior to transmission. It inserts a set of symbols referred to as a cyclic prefix (CP) in order to provide a guard time to prevent inter-block interference (IBI) due to multipath propagation. The transmitter also performs a linear filtering operation referred to as pulse shaping in order to reduce out-of-band signal energy.

Thus, transmitted data propagating through the channel can be modelled as a circular convolution between the channel impulse response and the transmitted data block, which in the frequency domain is a point wise multiplication of the DFT frequency samples. Then, to remove the channel distortion,the DFT of the received signal can simply be divided by the DFT of the channel impulse response.

1.7 Inter - Symbol Interference(ISI)

Inter-Symbol Interference (ISI) is a form of distortion of a signal in which one symbol interferes with subsequent symbols. This is an unwanted phenomenon as the previous symbols have similar effect as noise, thus making the communication less reliable. It is usually caused by multipath propagation or the inherent non - linear frequency response of a channel causing successive symbols to blur together. The presence of ISI in the system introduces error in the decision device at the receiver output. Therefore, in the design of the transmitting and receiving filters, the objective is to minimize the effects of ISI and thereby deliver the digital data to its destination with the smallest possibleerror rate.

1.8 Inter - Carrier Interference

Presence of Doppler shifts and frequency and phase offsets in an OFDM system causes loss in orthogonality of the sub-carriers. As a result, interference is observed between sub-carriers. This phenomenon is known as inter - carrier interference (ICI).

1.9 Understanding Concept of Cyclic Prefix

Cyclic Prefix can be best understood with the following example.Suppose you are driving a car in rain, and the car in front of you splashes a bunch of water on you. What do you do? You move further back and put a little distance between you and the front car, far enough so that the splash won't reach you. If we compare the reach of splash to delay spread of a splashed signal then we have a better picture of the phenomena and how to avoid it.[33]

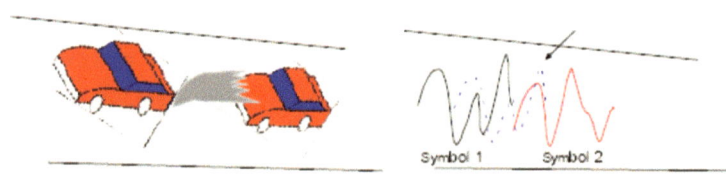

Figure 1.9 (a) Example of Delay spread

Increasing the distance avoid splashesfrom front car. The time elapsed to reach the splash is same as the delay spread of a signal. Figure 1.9(a)shows the symbol and its splash. In composite, these splashes are noise and affect the beginning of the next symbol.

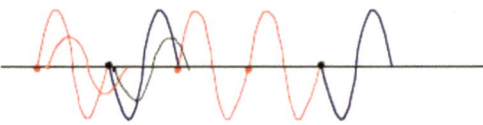

Figure 1.9 (b) Delayed version of the copied signal

To mitigate this noise at the front of the symbol, symbol is further moved away from the region of delay spread as shown below. A little bit of blank space has been added between symbols to catch the delay spread

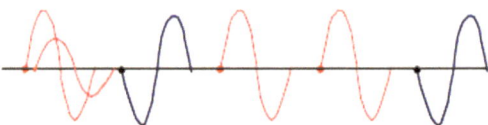

Figure 1.9 (c) Arrived delayed signal

But signalscannot have blank spaces. This will not work for the hardware which likes to crank out signals continuously. So there should be something here. Suppose the symbol run longer as the first choice.

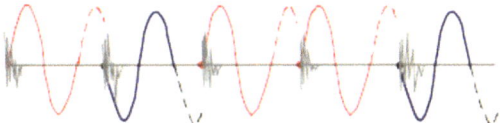

Figure 1.9 (d) Extended symbol using cyclic prefix

Suppose the symbol is extended into the empty space, so that the actual symbol is more than one cycle. But now the start of the symbol falls into the danger zone, and this start is the most important thing about the symbol because the slicer needs it to take a decision about the bit. The start of the symbol is undesired to fall in this region, so the symbol is slided backwards, so that the start of the original symbol lands at the outside of this zone.

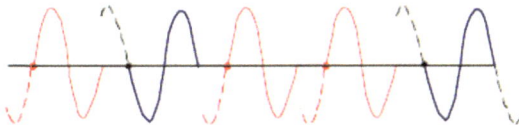

Figure 1.9 (e) Continuous signal after addition of cyclic prefix

1. The start of the symbol should be out of the delay spread zone so that it will not corrupt.
2. The signal should be started at the new boundary so that the actual symbol falls outside this zone.

  Therefore symbol is slided to start at the edge of the delay spread time and then the guard space is filled with a copy of tail end of the symbol. Then, the symbol is extended to 1.25 times long.To do this, copy the back of the symbol is taken and glued in the front. In reality, the symbol source is continuous, so the starting phase is adjusted ,making the symbol period longer.

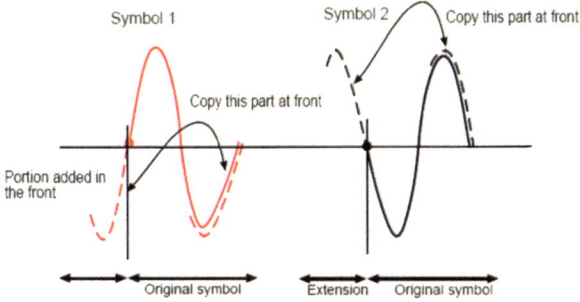

Figure 1.9 (f) Cyclic prefix of the signal

This procedure is called adding a cyclic prefix. The prefix is 10% to 25% of the symbol time. We add the prefix just after doing the IFFT. When the signal arrives at the receiver, this cyclic prefix is removed, to get back the original periodic signal. After doing FFT the symbols are recovered on each carrier. [13]

The Cyclic Prefix or Guard Interval is a periodic extension of the last part of an OFDM symbol that is added to the front of the symbol in the transmitter, and is removed at the receiver before demodulation.

The cyclic prefix has to two important benefits:

- The cyclic prefix acts as a guard interval. It eliminates the inter-symbol interference from the previous symbol.
- It acts as a repetition of the end of the symbol thus allowing the linear convolution of a frequency-selective multipath channel to be modeled as circular convolution which in turn maybe transformed to the frequency domain using a discrete fourier transform. This approach allows for simple frequency-domain processing such as channel estimation and equalization.

1.10 OFDM using Inverse DFT

Consider a data sequence $d_0, d_1, \ldots, d_{N-1}$, where each d_n is a complex symbol.(the data sequence could be the output of a complex digital modulator, such as QAM, PSK etc). Suppose we perform an IDFT on the sequence $2d_n$ (the factor 2 is used purely for scaling purposes), we get a result of N complex numbers S_m, $[m = 0,1,2, \ldots N-1]$ as:

$$S_m = 2\sum_{n=0}^{N-1} d_n \exp\left(j2\pi \frac{nm}{N}\right) = 2\sum_{n=0}^{N-1} d_n \exp(j2\pi f_n t_m) \quad (1.10.1)$$

Where, $[m = 0,1,2, \ldots N-1]$, $f_n = \frac{n}{NT}$ and $t = mT_s$

T_s represents the symbol interval of the original symbols. Passing the real part of the symbol sequence represented by equation (1.10.1) thorough a low-pass filter with each symbol separated by duration of T_s second yields the signal,

$$y(t) = 2Re\left\{\sum_{n=0}^{N-1} d_n \exp\left(j2\pi \frac{n}{T}\right)\right\}, for\ 0 \le t \le T \quad (1.10.2)$$

Where, T is defined as NT_s. The signal $y(t)$ represents the baseband version of the OFDM.

1.11 Advantages of OFDM

The Orthogonal Frequency Division Multiplexing (OFDM) transmission scheme has the following key advantages [34]

- OFDM is computationally efficient by using FFT techniques to implement the modulation and demodulation functions.
- By dividing the channel into narrowband flat fading sub channels, OFDM is more resistant to frequency selective fading than single carrier systems.
- By using adequate channel coding and interleaving, the symbols lost can be recovered, due to the frequency selectivity of the channel.
- OFDM is a bandwidth efficient modulation scheme and has the advantage of mitigating ISI in frequency selective fading channels.
- Channel equalization becomes simpler as compared to adaptive equalization techniques with single carrier systems.
- In conjunction with differential modulation, there is no need to implement a channel estimator.
- OFDM provides good protection against co-channel interference and impulsive parasitic noise.
- OFDM can easily adapt to severe channel conditions without complex time-domain equalization.
- OFDM eliminates Inter Symbol Interference (ISI) through the use of a cyclic prefix.
- OFDM is less sensitive to sample timing offsets than the single carrier systems.
- OFDM provides greater immunity to multipath fading and impulse noise.
- OFDM makes efficient use of the spectrum by allowing overlapping.

1.12 Disadvantages of OFDM

The Orthogonal Frequency Division Multiplexing (OFDM) transmission scheme is an attractive technology but has the following disadvantages:

- OFDM is more sensitive to carrier frequency offset and drift than single carrier systems, due to leakage of the Discrete Fourier Transform (DFT).
- OFDM is sensitive to frequency synchronization problems.
- OFDM is sensitive to Doppler Shift.
- The OFDM signal has amplitude with a very large dynamic range; therefore it requires

RF power amplifiers with a high Peak-to-Average Power Ratio (PAPR).
- The high PAPR increases the complexity of the Analog-to-Digital (A/D) and Digital-to-Analog (D/A) converters.
- The high PAPR also lowers the efficiency of power amplifiers.[34]

1.13 Peak to Average Power Ratio

Presence of large number of independently modulated sub-carriers in an OFDM system the peak value of the system, can be very high as compared to the average of the whole system. This ratio of the peak to average power value is termed as Peak-to- Average Power Ratio. The coherent addition of N signals of same phase produces a peak which is N times the average signal. The major disadvantages of a high PAPR is increased complexity in the analog to digital and digital to analog converter.

The peak to average power ratio for a signal is defined as:

$$PAPR = \frac{\max[x(t)x^*(t)]}{avg\ [x(t)x^*(t)]} \quad (1.13.1)$$

Where $x^*(t)$ corresponds to the conjugate operator, max $[x(t)x*(t)]$ is peak value of the signal, $avg[x(t)x*(t)]$ is mean square value of the signal.[20]

Expressing in decibels

$$PAPR_{dB} = 10log_{10}(PAPR) \quad (1.13.2)$$

1.14 PAPR Reduction Techniques

The high Peak-to-Average Power Ratio (PAPR) or Peak-to-Average Ratio (PAR) or Crest Factor of the Orthogonal Frequency Division Multiplexing (OFDM) systems can be reduced by using various PAPR reduction techniques as follows [20]:

- ❖ Tone Reservation (TR).
- ❖ Clipping.
- ❖ Companding Transforms.
- ❖ Constellation Shaping.
- ❖ Phase Optimization.
- ❖ Tone Injection (TI).
- ❖ Block Coding.
- ❖ Partial Transmit Sequence (PTS).

- ❖ Selective Mapping (SLM).
- ❖ Interleaving.
- ❖ Active Constellation Extension Methods.

CHAPTER 2

Literature Review

2.1 Different methods for Peak-to-Average Power (PAPR) Reduction in Orthogonal Frequency Division Multiplexing (OFDM)

Himanshu Bhushan Mishra et al.in 2012 proposed a new Selective-Mapping (SLM) technique in WIMAX without side information which is the major issue in theclassical SLM Technique. In this paper the PAPR performance is measured using complementary cumulative distribution function (CCDF) plot and the probability of SI detection error performance have been evaluated as the criteria for WiMAX standard IEEE 802.16e. WiMAX with its standard IEEE 802.16d/e is the advanced technology used for long range communication with high data rate. It is well known that the Orthogonal Frequency Division Multiplexing (OFDM) is a promising technique for getting high data rates in a multipath fading environment. Hence, the physical layer of WiMAX uses OFDM. But the main disadvantage of OFDM is the high peak to average power ratio (PAPR). In this paper PAPR reduction is achieved using selected mapping (SLM) technique and simultaneously without sending the side information (SI) along with the OFDM symbol.[1]

E. Al-Dalakta et al. in 2012 proposed an efficient technique for reducing the biterror rate (BER) of Orthogonal FrequencyDivision Multiplexing (OFDM)signals transmitted over nonlinear solid-state power amplifiers (SSPAs).The proposed technique is based on predicting the distortion power thatan SSPA would generate due to the nonlinear characteristics of suchdevices. Similar to the Selective-Mapping (SLM) or Partial-Transmit-Sequence(PTS) schemes, the predicted distortion is used to select a set of phasesthat minimize the actual SSPA distortion. Simulation results confirmedthat the signal-to-noise ratio that is required to obtain a BER of $\sim 10^{-4}$ using the proposed technique is less by about 8 dB when it is compared tothe standard PTS utilizing 16 partitions. Moreover, complexity analysisdemonstrated that the proposed system offers a significant complexityreduction of about 60% compared to state-of-the-art methods.This work demonstrated that less direct PAPR indicators can provide better performance when combined with distortionless techniques such as PTS and SLM. Therefore, the proposed techniques are optimized to combat the consequences of high PAPR rather than reducing the PAPR itself. The proposed techniques are based on using the distortion level to select the optimal PTS and SLM system parameters. [2]

Shiann-Shiun Jeng et al. in 2011 proposes a new method based on companding Peak-to-Average Power Ratio (PAPR) Reduction of Orthogonal Frequency Division Multiplexing (OFDM) signals. This paper suggests that uniformly distributed companding scheme and piecewise companding scheme cannot deliver the performance that satisfies various requirements of the system. So, the distribution of the OFDM signal is transformed into the trapezium distributionand the general formulas for the proposed scheme are derived that enable the de- sired performance to be achieved by controlling the parameter.The simulation results reveal that the proposed scheme may offer the more efficient PAPR reduction or the lower BER than the uniformly-distributed and piecewise schemes under the condition of efficient PAPR reduction or efficient BER performance. [3]

Suma M N et al. gives a survey on developments in OFDM so far. He suggested that apart from high Peak-to-Average Power Ratio (PAPR) there are many more techniques that are needed to increase the performance of OFDM systems like interference cancellation, phase noise mitigation, Synchronization among carriers and post equalization. In today's communication scenario, high data rate single-carrier transmission may not be feasible due to too much complexity of the equalizer in the receiver. To overcome the frequency selectivity of the wide band channel experienced by single-carrier transmission, multiple carriers can be used for high rate data transmission. Orthogonal Frequency Division Multiplexing (OFDM), is multicarrier system which has become a modulation in physical layer of next generation WiMAX, LTE system. In this work effort is made to present challenges in OFDM and work done so far in channel equalization and different transforms used in OFDM system like Discrete Fourier Transform (DFT) Based OFDM, Discrete Cosine Transform (DCT) Based OFDM, Wavelet based and Wavelet packet based OFDM, Discrete Hartley Transform (DHT) Based OFDM, and Coded OFDM which includes Turbo codes and Alamouti codes. [4]

Jun Hou et al. proposed a nonlinear companding scheme toreduce the Peak-to-Average Power Ratio (PAPR) and improve Bit Error Rate (BER) for OFDMsystems. This proposed scheme mainly focuses on compressingthe large signals, while maintaining the average power constantby properly choosing transform parameters. Moreover, analysis shows that the proposed scheme without decompanding atthe receiver can also offer a good BER performance. Finally,simulation results show that the proposed scheme outperformsother companding scheme in terms of spectrum side-lobes, PAPRreduction and BER performance. [5]

Yuan Jiang in 2010 suggest a new companding algorithmthat offers improved performance in terms ofBER and OBI while reducing PAPR effectively.Orthogonal Frequency Division Multiplexing (OFDM) mitigates the effect of intersymbol interference (ISI) but it suffers from inter-block interference (IBI).A good remedy for the OBI is companding. This paper proposes and evaluates a new compandingalgorithm. This work uses the special airy function and isable to offer an improved bit error rate (BER) and minimizedOBI while reducing PAPR effectively. [6]

Kitaek Bae et al. in 2010 proposed a new method for Peak-to-Average Power Ratio (PAPR) Reduction in Orthogonal Frequency Division Multiplexing (OFDM), a novelActive Constellation Extension (ACE) algorithm with adaptive clipping control namely Adaptive ACE which is an improvement in the technique based on the combination of two techniques namely clipping and Active Constellation Extension (ACE) known as Clipping-Based Active Constellation Extension (CB-ACE) technique. This work suggests that CB-ACE cannot achieve the minimum PAR when the target clippinglevel is set below an initially unknown optimum value. Toovercome this low clipping ratio problem, Adaptive ACE is proposed. Simulation resultsdemonstrate that proposed algorithm can reach the minimumPAR for severely low clipping ratios. In addition, thetradeoff between PAR and the loss in E_b/N_o over an AWGNchannel in terms of the clipping ratio has also been described. [7]

Ms. V. B. Malode et al. in 2010 proposed a new method to reduce Peak-to-Average Power Ratio (PAPR) Peak-to-Average Power Ratio (PAPR) by probabilistic method,modified selective mapping technique using the standard arrays of linear block codes. In this work lowest PAPR in each coset of a linear block codes is chosen as its coset leader from several transmitted signal , which further results in high performance ofwireless communication.[8]

Stephane Y. Le Goff et al. in 2009 suggest an improvement in classical Selective Mapping (SLM) technique and proposed a new method in which SLM can be implemented without sending the side information which is a major issue in classical Selective Mapping (SLM) technique .SLM requiresthe transmission of several side information bits for each datablock, which results in some data rate loss. These bits mustgenerally be channel-encoded because they are particularlycritical to the error performance of the system. This increases thesystem complexity and transmission delay, and decreases the datarate even further. In this paper, we propose a novel SLM methodfor which no side information needs to be sent. This technique is particularly attractive for systemsusing a large number of subcarriers and the probabilityof SI detection error can be made very small by increasingthe extension factor and/or the number of subcarriers. [9]

Sulaiman A. Aburakhia et al. in 2009 proposed a new linear companding transform (LCT)withmore design flexibility than linear nonsymmetricalcompanding transform (LNST) .Simulation results have been shown comparing both the above technique assuming AWGN channel .The results show that the proposed method has a higher PAPR reduction capability and better BER performance than LNST, with less spectral broadening. This work suggests that the proposed can be designed to meet system requirements, power amplifier characteristics, and achieve an excellent tradeoff betweenPAPR reduction and BER performance. Furthermore, the proposed transform is simple to implement and has no limitationson the system parameters such as number of subcarriers modulation order, or constellation type. [10]

F.S. Al-kamaliet al. suggested the design of a newtransceiver scheme for the SC-FDMA scheme using thewavelet transform. No redundancy is added to the newsystem because of the discrete wavelet transform. Thus, itscomplexity is slightly increased as compared to theconventional SC-FDMA scheme.This workdescribes that the proposed scheme has better PAPR performance and BER performance than theconventional SC-FDMA scheme.Theproposed HW-SC-FDMA scheme provides about 3 dBgain when compared to the conventional SC-FDMAscheme over the vehicular A channel.[11]

Tao Jiang et al. in 2008 gives a reviewand analysis different OFDM PAPR reduction techniques, basedon computational complexity, bandwidth expansion, spectralspillage and performance discussing some methods of PAPRreduction for multiuser OFDM broadband communication systems. This paper clarifies the definition of PAPR for baseband PAPR and passband PAPR and deduces that the passband PAPR is approximately twice thebaseband PAPR.This paper also highlights the motivation of PAPR Reduction.[12]

Satoshi Kimuraet al.proposes a deep clipping method to suppress the peakregrowth and to reduce the Peak-to-Average power Ratio (PAR)of OFDM signals without iteration. The mainidea of deep clipping is modifying the clipping function todeeply clip the high amplitude peaks to a level smaller than aclipping level. This proposed method achieves significant reduction of thepeak regrowth that is equivalent to that by four times repeated Clipping and Filtering (CAF) method. Meanwhile,in-band distortion becomes larger than that of the four timesrepeated CAF.[13]

Orlandos Grigoriadis et al. in 2008 gives use a Matlab simulation of OFDM to see how the Bit Error Rate (BER) of a transmission varies when Signal to Noise Ratio (SNR) and Multi-propagation effects are changed on transmission channel.Moreover, some of the main variables of the code are described, as the choice of them has a critical effect on the results.This work also suggests that if there is any increase in the number of carriers for certain SNR, multipropagation effect also increases thereby resulting in the increase in BER.[14]

Josef Urban et al. in 2007 proposed a combination of Interleaving and repeated Clipping and Filtering to reduce the Peak-to-Average power Ratio (PAR)of Orthogonal Frequency Division Multiplexing (OFDM)signals in order to increase the overall performance for the PAPR reduction. The performance is evaluated in AWGN channel with presence of Saleh nonlinearity model. Another contribution of the paper is a study of influence of side information coding on total bit error probability.Thiswork also deals with side information needed for interleaving method and the influence of its representationon bit error rate.The main advantage of the proposed combination lies in substantial BER reduction in AWGN and nonlinearchannel. The disadvantage of the method is a need for side information transmission.[15]

Robert J. Baxleyet al. in 2007 suggest a comparison between two selected mapping (SLM) and partial transmit sequence (PTS)fororthogonal frequency division multiplexing (OFDM). This work also suggests thatthe overall computational complexity of PTS is lower thanthat of SLM in certain cases, and that SLM always has betterPAR reduction performance than PTS.Thetwo schemes are compared using three different performance metrics by assuming a given amountof computational complexity.This work concludes that SLM is preferred to PTS because it is conceptually simpler, does not require any off-line complexityoptimization with respect to as would be recommended forPTS and performs better than optimized PTS.[16]

Hyung G. Myunget al. in 2006 suggested a review of Single Carrier Frequency Division Multiplexing (SC-FDMA) and Orthogonal Frequency Division Multiplexing (OFDM).This paper describes that SC-FDMA hassimilar throughput performance and thesame overall complexity as OFDMA. A principaladvantage of SC-FDMA is the peak-to-average powerratio (PAPR), which is lower than that of OFDMA. SC-FDMA is currently a strong candidate for the uplink multiple access scheme in the Long Term Evolution of cellularsystems under consideration by the Third Generation Partnership Project (3GPP).This paper highlightssome of the important issues in the design of a SC-FDMAsystem and compares LFDMA and IFDMAwith respect to the two major performance indicators,system throughput and PAPR. For each configuration, throughput measures with static subcarrierassignments and with channel dependent scheduling is presented.[17]

Dov Wulichin in 2005 suggested concept of efficient Peak- to-average Power Reduction (PAPR) as all the techniques used to reduce the same concentrate only on the reduction of PAPR without taking into account various important parameters like BER increase and power increase in orthogonal frequency division multiplexing (OFDM) signal.The effect of PAPR on BER is considered when theefficiency-PAPR relationship of the power amplifier is takeninto consideration. It is shown that a PAPR may exist forwhich the BER reaches a minimal value. [18]

Seung Hee Han et al. in 2005 proposed an overview that describes some of the important on Peak- to-average Power Reduction (PAPR) techniques for multicarrier transmission including amplitude clipping and filtering,coding, partial transmit sequence, selected mapping, inter-leaving, tone reservation, tone injection, and active constellation extension. This paper also describes comparison of above mentioned techniques on the various parameters such as complexity, distortion introduced, power increase and data rate loss which is also the criteria for the selection of Peak- to-average Power Reduction (PAPR) techniques for various applications. No specific PAPR reduction technique is thebest solution for all multicarrier transmissionsystems. Rather, the PAPR reduction techniqueshould be carefully chosen according to varioussystem requirements. In practice, the effect ofthe transmit filter, D/A converter, and transmitpower amplifier must be taken into consideration to choose an appropriate PAPR reductiontechnique. This works also gives a clarification on how Peak- to-average Power Reduction (PAPR) techniques can be implemented on Orthogonal Frequency Division Multiple Access (OFDMA) and Multiple Input Multiple Output-Orthogonal Frequency Division Multiplexing (MIMO-OFDM). [19]

Tao Jiang et al. in 2005 proposed a new nonlinear companding technique, called exponential companding, to reducethe high Peak-to-Average Power Ratio (PAPR) of Orthogonal Frequency Division Multiplexing (OFDM) signals. Unlike the μ-lawcompanding scheme, which enlarges only small signals so thatincreases the average power, the schemes based on exponentialcompanding technique adjust both large and small signals andcan keep the average power at the same level. By transformingthe original OFDM signals into uniformly distributed signals(with a specific degree), the exponential companding schemescan effectively reduce PAPR for different modulation formatsand sub-carrier sizes. Moreover, many PAPR reduction schemes,such as µ -law companding scheme, cause spectrum side-lobesgeneration, but the exponential companding schemes cause less spectrum side-lobes. This work suggest that the proposed exponential companding schemes canoffer better PAPR reduction, Bit Error Rate (BER), and phaseerror performance than the µ -law companding scheme.[20]

J. Armstrong in 2002 proposed a new namely repeated clipping and frequency domain filtering (RCF). It is shown that repeated clipping and frequency domain filtering of an orthogonal frequency division multiplexing (OFDM) signal can significantly reduce the peak-to-average power ratio (PAPR) of the transmitted signal. The technique causes no increase in out-of-band power. Significant PAPR reduction can be achieved with only moderate levels of clipping noise. This work suggest that the PAPR of an OFDM signal can be reduced without any increase in the out-of-band power by clipping the oversampled time domain signal followed by filtering using an FFT-based,frequency domain filter designed to reject out-of-band discrete frequency components. Filtering results in peak regrowth. Further, PAPR reduction can be achieved by repeated clipping and filtering operations. The distortion of the in-band signal results in shrinking of the overall signal constellation and an added noise-like effect.[21]

J. Armstrong in 2001, new PAPR reduction technique has been described in which an interpolated version of the baseband signal is clipped and then filtered with a new form of filter. The filter consists of a forward and an inverse FFT which is designed to remove the out-of-band noise without distorting the in-band discrete signal. It is shown that significant PAPR reduction can be achieved without any increase in out-of-band power. Some in-band distortion results but this will have negligible effect on the overall BER in most systems. The distortion has two effects: an overall shrinking of the constellation, which isautomatically corrected by the receiver AGC and a noise like component. The clipping noise is added at thetransmitter rather than the receiver and so fades along withthe signal in a fading channel. This work presents the combination of the system with non-ideal amplifiers. It is shown that the level of out-of-band power depends on both the degree of PAPR reduction and on the linearity of the amplifier. The new technique allows simple trade-offs between in-band distortion, amplifier back-off and amplifier linearity. The new technique is completely compatible with other aspects of transmitter design such as windowing and filtering. It can be implemented by replacing the transmitter IFFT with an oversize IFFT, followed by the clipping and filtering circuit. The technique could also be viewed as a way of making slight changes to the input data vector in a way that reduces the PAPR. No changes are required in the receiver so can be adopted without anychange to telecommunications standards.[22]

D. S. Jayalathet al. in 2000 presents a simplified version of partial transmit sequences (PTS) and a new scheme for selected mapping sequences to reduce peak-to-average power ratio (PAP) of an OFDM signal. Simplification of PTS is achieved by having a set of partitions but optimizing phase values only for alternate partitions.This proves to be a promising solution to reduce complexity of PTS. It is also proposed to choose the selected mapping sequences are using Newman phase sequences. In this case, albeit with an increase of the complexity, very high PAR reduction can be achieved.[23]

Xiaodong Liet al. in 1998 investigated the effects of clipping andfiltering on the performance of orthogonal frequency division multiplexing (OFDM).With a clipping ratio around 1.4 and filtering, the CF of abandpass OFDM signal with 128 tones at the 99.999% pointis reduced from 13 to about 9 dB, which is comparable to theabsolute CF of a raised-cosine pulse-shaped bandpass QPSKsignal.[24]

R. W. Bauml et al. in 1996 proposed a new method namely Selective Mapping (SLM) to reduce the PAPR by which significant gains can be achieved by selected mapping whereas complexity remains quite moderate. This proposed method can be used for variety of applications and can be used for arbitrary numbers of carriers and any signal constellation. This work also suggests that even in single carrier systems where PAR grows as the roll off factor of the pulse shaping filter decreases, selected mapping can be applied advantageously.[25]

CHAPTER 3

PROBLEM IDENTIFICATION

In a telecommunications context, there are many instances of nonlinearities, among them quantization, digital-to analog conversion, the propagation channel, the low-noise amplifier, hard decision, etc. In the context of this work, the focus is on the power amplifier.[34]

Figure 3.1 shows a typical class A amplifier characteristics. The amplitude AM/AM characteristic curve relating the input and output powers, the amplifier gain, and the efficiency. The efficiency η is maximum at the saturation power and drops if lower input powers are considered. [34]

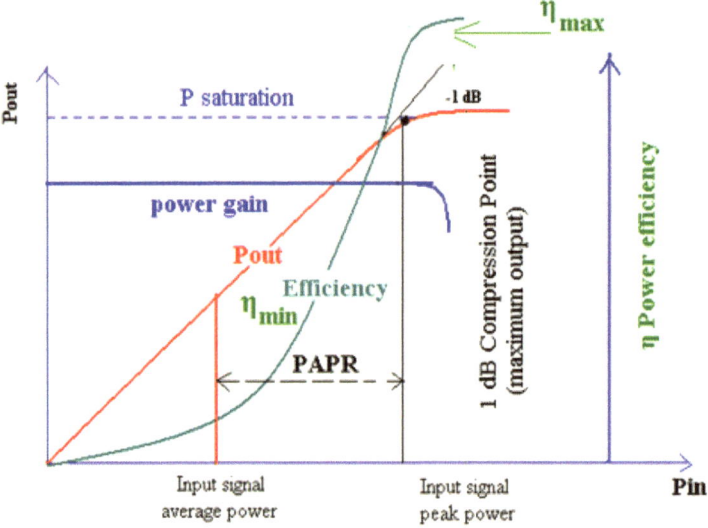

Figure 3.1. Power amplifier characteristics

The power amplifier characteristicsis illustrated in Figure 3.1.If the maximum input power is equal to the saturation, it can be easily understood that small reduction in the PAPR can permit amplification close to the saturated power, with a consequent increase in efficiency. On the other hand, a high PAPR implies that the mean input power falls in the linear region, where the efficiency is lower. [34]

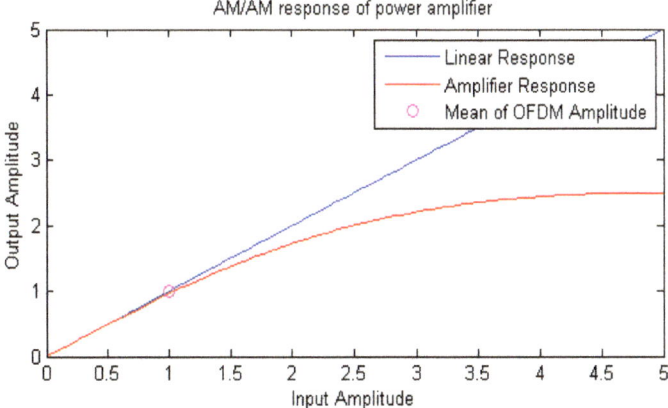

Figure 3.2. Amplifier Response of OFDM signal

Ideal power amplifiers are perfectly linear up to a saturated power called P_{sat}. Within this region, the output power is equal to the input power multiplied by the amplifier gain. However, practical power amplifiers are not linear, and the output power will be a function of the input power according to this nonlinear power amplifier characteristics. These nonlinear relations imply nonlinear distortions, creating in and out-of-band noise, spectral regrowth, nonlinear intersymbol interference, and hence, degradation of bit error rates. For class A power amplifiers, to limit these drawbacks, the input power must be driven within the linear region (i.e., backed off from the saturated power), which improves linearity but reduces power efficiency and increases heat dissipation. Therefore, linearity is typically achieved either by reducing efficiency or by using linearization techniques.[34].

Figure 3.3 and Figure 3.4 gives the non-linearities caused by the OFDM signal after passing through High Power Amplifier. To overcome the problem of high PAPR the dynamic range of the power amplifier can be increased to accommodate the highest amplitude value of the envelope within the linear range of the amplifier. But, higher dynamic range implies higher cost of the power amplifier. Also, with larger dynamic range, heat loss is more. This reduces the efficiency of the Power Amplifier.

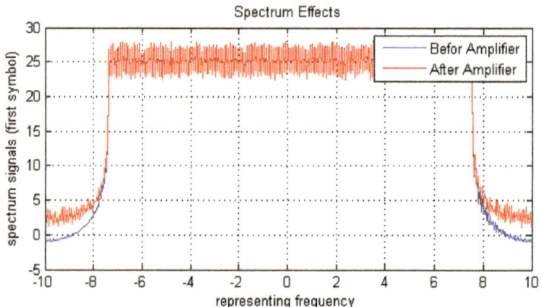

Figure 3.3 Distortion in the Spectrum of OFDM Signal after passing through HPA.

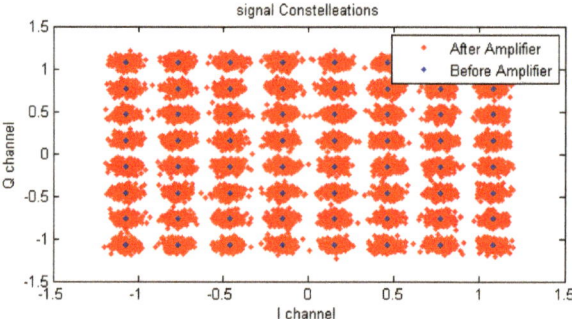

Figure 3.4 Distortion in the Signal Constellation of OFDM Signal after passing through HPA.

Another way to overcome PAPR is that the peak transmit power can be limited by either regulatory or application constraints. If the peak transmit power is limited by either regulatory or application constraints, the effect is to reduce the average power allowed under multicarrier transmission relative to that under constant power modulation techniques. This in turn reduces the range of multicarrier transmission.So, the several techniques to reduce the PAPR of the generated OFDM signal can be used, so that before transmitting, the PAPR of the signal is lowered to an acceptable limit.

There is a unique feature of implementing and finding out facts about PAPR reduction techniques, discussing it's advantages and disadvantages, recommending designs to overcome the problem of PAPR in OFDM. Since the work on comparative study of PAPR reduction techniques on account of different number of sub-carrier used in OFDM signal is less so, this work reflects comparison of some of the techniques and an improved technique has been proposed for this system.

To compare the PAPR reduction techniques, simulation results of each reduction technique are calculated and observed. Before deciding specified method for comparison of PAPR reduction technique in OFDM , it is important to understand the various options available to us. Now we will go through the various PAPR reduction techniques as follows.

Two kinds of approaches are investigated which assure that the transmitted OFDM signal does not exceed the amplitude threshold which is set according to the input back-off.

- The first approach makes use of redundancy in such a way that any data sequence leads to the magnitude of OFDM signal greater than A_0 or that at least the probability of higher amplitude peaks is greatly reduced. This approach does not result in interference of the OFDM signal.
- In the second approach, the OFDM signal is manipulated by a correcting function, which eliminates the amplitude peaks. The out of band interference caused by the correcting function is zero or negligible. However, interference of the OFDM signal itself is tolerated to a certain extent.

3.1 Clipping and Filtering

The simplest and most widely used technique of PAPR reduction is to basically clip the parts of the signals that are outside the allowed region [22]. For example, using HPA with saturation level below the signal span will automatically cause the signal to be clipped. For amplitude clipping, that is

$$C(x) = \begin{cases} x, & |x| \leq A \\ A, & |x| > A \end{cases}$$

where A is preset clipping level and it is a positive real number. Generally, clipping is performed at the transmitter. However, the receiver need to estimate the clipping that has occurred and to compensate the received OFDM symbol accordingly. Typically, at most one clipping occurs per OFDM symbol, and thus the receiver has to estimate two parameters: location and size of the clip. However, it is difficult to get these information. Therefore, clipping method introduces both in band distortion and out of band radiation into OFDM signals, which degrades the system performance including BER and spectral efficiency.

Filtering can reduce out of band radiation after clipping, although it cannot reduce in-band distortion. However, clipping may cause some peak regrowth, so that the signal after clipping and filtering will exceed the clipping level at some points. To reduce peak regrowth, a repeated clipping-and-filtering operation can be used to obtain a desirable PAPR at a cost of computational complexity increase. As improved clipping methods, peak windowing schemes attempt to minimize the out of band radiation by using narrowband windows .

3.2 Coding

Coding is another method to reduce the PAPR. A simple idea introduced in [22] is to select those codewords that minimize or reduce the PAPRfor transmission. This idea is illustrated in the following example.

Table 3.2 PAPR values of all possible data blocks for an OFDM signal with four subcarriers and BPSK modulation.

Data Block X	PAPR (dB)	Data Block X	PAPR (dB)
$[1,1,1,1]^T$	6.0	$[1,1,1,1]^T$	2.3
$[1,1,1,-1]^T$	2.3	$[1,1,1,1]^T$	3.7
$[1,1,-1,1]^T$	3.7	$[1,1,1,1]^T$	6.0
$[1,1,-1,-1]^T$	3.7	$[1,1,-1,-1]^T$	2.3
$[1,-1,1,1]^T$	2.3	$[-1,-1,1,1]^T$	3.7
$[1,-1,1,-1]^T$	6.0	$[-1,-1,1,-1]^T$	2.3
$[1,-1,-1,1]^T$	3.7	$[-1,-1,-1,1]^T$	2.3
$[1-,1,-1,-1]^T$	2.3	$[-1,-1,-1,-1]^T$	6.0

Example: The PAPR for all possible data blocks for an OFDM signal with four subcarriersand binary phase shift keying (BPSK) modulation is shown in Table 3.2. It can be seen from this table that four data blocks result in a PAPR of 6.0 dB, and another four data blocks result in a PAPR of 3.7 dB. It is clear that we could reduce PAPR by avoiding transmitting those sequences. The above mentioned technique can be applied by block coding the data such that the 3-bit data word is mapped onto a 4-bit codeword such that the set of permissible sequences does not contain those that result in high PAPR. The PAPR of the resulting signal is 2.3 dB, a reduction of 3.7 dB from that without block coding.

However, this approach suffers from the need to perform an exhaustive search to find the bestcodes and to store large lookup tables for encoding and decoding, especially for a large number of subcarriers. Moreover, this approach does not address the problem of error correction.

3.3 Interleaving

Interleaving method [12], sometimes named as Adaptive symbol selection method, is based on creation of multiple OFDM signals by bit interleaving of input bit sequence. One of the simplest ways to perform the interleaving is the use of matrix interleaver in which data are written by lines and read by columns.

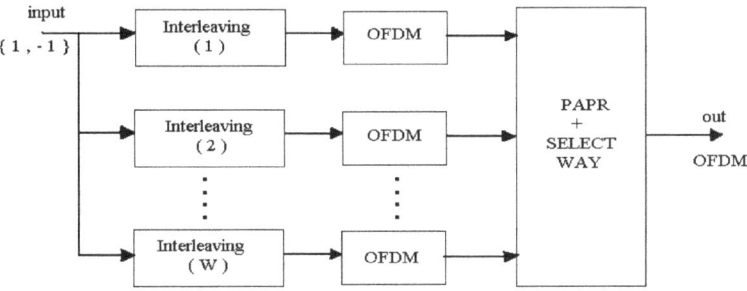

Figure 3.3. Principle Schematic of PAPR Reduction by Interleaving [12]

The PAPR reduction method based on interleaving use W ways, each of them using different interleaving matrix. After each interleaver, the OFDM modulation is performed using IFFT. Finally, the way with the lowest PAPR is chosen from all W realizations.

The only parameter that influences the PAPR reduction performance is the number of interleavers (ways) W used .As the random character of the input bit stream is expected , the amount of PAPR Reduction does not depend on the exact parameters of matrices to interleave .In the receiver, the matrics transposed to the matrix selected in the transmitter is used to de-interleave .[12]

3.4 Companding

In this method we compress the dynamic range of the signal by a memory-less transformation at the transmitter and expand the amplitude level at the receiver by considering the approximate Rayleigh distribution of the OFDM amplitudes. This transformation essentially changes the probability distribution of the amplitude of OFDM signal and achieves the PAPR reduction by both enlarging the small amplitudes and compressing large signals. The power is adaptively allocated for each sub-carrier according to the distribution in each block. However, the average signal power increases after the compression and the compressed signal still exhibit non-uniform distributions.

The companding transform scheme uses a compander to reduce the signal amplitude. Such an approach can effectively reduce the PAPR with a less computational complexity . The corresponding transmitter and receiver need a compander and an expander, respectively, which of course slightly increases the hardware cost. [12]

3.5 Peak Windowing

The simplest way to reduce the PAPR is to clip the signal, but this significantly increases the out of band radiation. Since the OFDM signal is multiplied with several of these windows the resulting spectrum is a convolution of the original OFDM spectrum with the spectrum of the applied window. So, ideally the window should be as narrow band as possible. On the other hand, the window should not be too long in the time domain, because that implies that many signal samples are affected, which increases the bit error ratio. Examples of suitable window functions are the Cosine, Kaiser and Hamming window. PAPR could be achieved, independent from number of sub-carriers, at the cost of a slight increase in BER and out of band radiation.

3.6 Additive Correcting Function

This approach is similar to peak windowing but the additive correction function is used instead of multiplicative correction function. OFDM signal is modified in such a way that a given amplitude threshold of the signal is not exceeded after the correction. With the restriction that the correcting function dose not causes any out of band interference, the correcting function has been given with minimal power and thus causes minimal interference power within the OFDM band.

3.7 Selected Mapping (SLM)

Bäuml, Fischer and Huber (1996) [26] proposed this method to reduce PAPR for a wide range of applications. Because of the statistical independence of the carriers, the corresponding time domain samples in the equivalent complex valued lowpass domains are approximately Gaussian distributed. This results in a high peak to average power ratio. Because of varying assignment of data to the transmit signal, this method is called "Selected Mapping". The core concept is to choose one particular signal, which exhibits some desired properties out of N signal representing the same information. Then all N frames are transformed into the time domain and the one with the lowest PAPR is selected for transmission. To recover data, the receiver has to know which vector has actually used and the number n of the vector is transmitted to the receiver as side in formation. This method can be used for arbitrary number of carriers and any signal constellation. It provides significant gain against moderate additional complexity.[26]

3.8 Tone Reservation

In this method, the basic idea is to reserve a small set of tones for PAPR reduction. Fortunately, the problem of computing the values for these reserved tones which minimize the PAPR can be formulated as a convex problem and can be solved analytically. The amount of PAPR reduction depends on the number of reserved tones, their locations within the frequency vector, and the amount of complexity. This method is effective when no. of total subcarriers is much larger than the reserved tones otherwise the throughput bit rate decreases considerably.[45][46]

3.9 Tone Injection

This method achieves PAPR reduction of multicarrier signal without data rate loss. The basic idea is to increase the constellation size, so that each of the points in the original basic constellation can be mapped into several equivalent points in the expanded constellation. If these duplicate signal points are spaced by $D = \rho \, d \, M$ where M is the constellation size, and $\rho \geq 1$, the BER will not increase and the only addition to the standard receiver is module-D addition after the FFT. Since each information unit can be mapped into several equivalent constellation points, these extra degrees of freedom can be exploited for PAPR reduction.[13]

3.10 Selective Scrambling (Interleaving)

Usually constellation points that lead to high-magnitude time signals are generated by correlated bit patterns (for example, a long string of ones or zeros). Therefore, by scrambling the input bit streams, we may reduce the probability of large peaks generated by those bit patterns. For example, the method is to form four code words in which the first two bits are 00, 01, 10 and 11 respectively. The message bits are first scrambled by four semi random interleaver with length N. Then the branch with the lowest PAPR is selected and one of the pair of bits defined earlier is added at the beginning of the selected sequence. At the receiver, these first two bits are used to select the suitable descrambler. PAPR is typically reduced while incurring negligible redundancy in a practical system. However, an error in the bits that encode the choice of scrambling sequence may lead to long propagation of decoding errors.

CHAPTER 4

METHODOLOGY

4.1 Objectives

The objectives of the proposed method is to reduce the Peak-to-Average Power Ratio (PAPR) in the Orthogonal Frequency Division Multiplexing (OFDM) systems are –

1. To study and evaluate the original OFDM signal.
2. To study and evaluate the existing technique Selective Mapping.
3. To study and evaluate the existing techniques Active Constellation Extension combined with Repeated Clipping and Filtering (RCF).
4. To study and evaluate the existing technique namely Companding Transform.
5. To improve existing technique Selective Mapping to get better performance.

4.2 Hardware and Software Required

4.2.1 Hardware Required

Processor	: Intel (R) Pentium (TM) 2 Dual Core
Processor Speed	: T7100 @1.80GHz
Operating System	: WXP Service Pack3
Hard Disk	: 120 Gb
RAM	: 2 Gb

4.2.2 Software Required

The software used for proposed project is MATLAB version 7.10.0.499 (R2010a). MATLAB (Matrix Laboratory), a proprietary product of Math Works, was created in the late 1970s by Cleve Moler, the chairman of the Computer Science department at the University of New Mexico. The Matrix Laboratory is a numerical computing environment and fourth generation programming language. The Matrix Laboratory is developed by Math Works. MATLAB is a high level language which allows matrix manipulations, plotting of functions and data, implementation of algorithms, creation of user interfaces and interfacing with programs written in other languages including C, C++ and Fortran. The MATLAB application is built around the MATLAB language. The simplest way to execute MATLAB code is to type it in the Command Window, which is one of the elements of the MATLAB Desktop. When the code is entered in the Command Window, the software can be used as an interactive mathematical shell. A sequence of commands can be saved in a text file named as the MATLAB Editor.

4.3 Simulation model of OFDM System

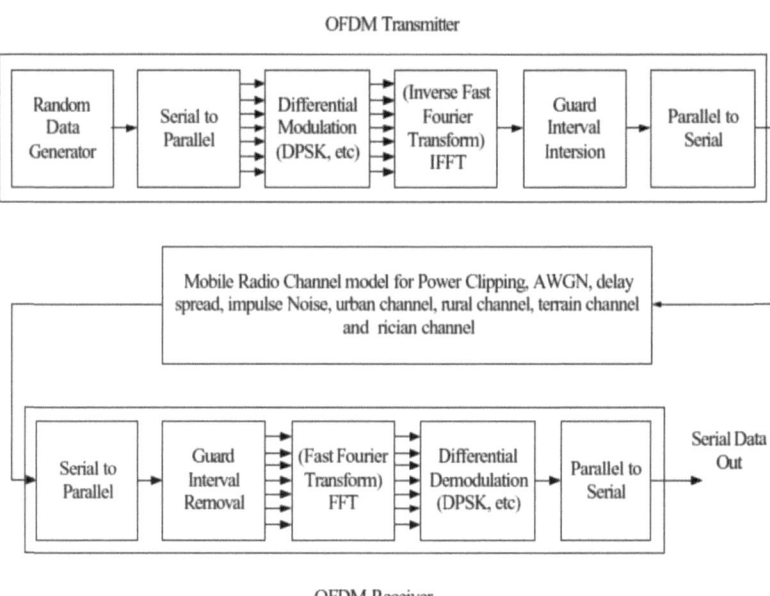

Figure 4.2.1 Block Diagram of OFDM System[36]

4.3.1 Random Data Generator

Random data generator is used to generate a serial random binary data. This binary data stream models the raw information that going to be transmitted. The serial binary data is then fed into OFDM transmitter.

4.3.2 Serial to Parallel Conversion

The input serial binary data stream is formatted into word size required for transmission, ex. 2 bit/words for QPSK, and shifted into parallel format. The parallel data is then transmitted in parallel by assigning each data word to one carrier in the transmission.

4.3.3 Modulation of Data

The data to be transmitted on each carrier is then differential encoded with previous symbols, then mapped into a Phase Shift Keying (PSK) format. Since differential encoding requires an initial phase reference, an extra symbol is added at the start for this purpose. The data on each symbol is then mapped to a phase angle based on the modulation method. For example, in the QPSK the phase angles used are 0, 90, 180, and 270 degrees. The use of phase shift keying produces a constant amplitude signal and was chosen for its simplicity and to reduce problems with amplitude fluctuations due to fading.

4.3.4 Inverse Fourier Transform

After the required spectrum is worked out, an Inverse Fourier Transform is used to find the corresponding time waveform. The guard period is then added to the start of each symbol.

4.3.5 Guard Period

The type of guard period used in this simulation was a cyclic extension of the symbol. The length of guard period is chosen as 25 percent of the length of symbol.

4.3.6 Parallel to Serial Converter

After guard period has been added, the symbol is then converted back to a serial time waveform. This signal is the baseband signal for the OFDM transmission.

4.3.7 Channel

A channel model is then applied to the transmitted signal. The effect of mobile radio channel impairments such as peak power clipping, AWGN, delay spread and impulse noise were simulated separately.

The power clipping is applied to the OFDM signal by cutting the signal that higher than a certain determined power value. The AWGN is applied to the OFDM signal by adding the AWGN with noise factor $10^{-(SNR/10)}$ to the transmitted signal. The delay spread added by simulating the delay spread using FIR filter. The length of FIR filter represents the maximum delay spread while the coefficient amplitude represent the reflected signal magnitude. The impulse noise is added to the OFDM signal by creating some short extreme noise that occurs repeatedly. The range between two impulse noises is set 3761 times of the length of impulse noise.

The urban channel, rural channel, terrain channel and rician channel can be simulated by using two Matlab Communication Toolbox, those were Multipath Rayleigh Fading Channel toolbox and Additive White Gaussian Noise Toolbox. The channels are differentiated by setting power profile delay and mobile velocity differently for each channel. The effect of mobile velocity to the performance of OFDM can be simulated by varying the maximum Doppler effect in Multipath Rayleigh Fading Channel Toolbox.

4.3.8 Receiver

The receiver basically does the reverse operation inverse to the transmitter. The guard period is removed and FFT of each symbol is then taken to find the original transmitted spectrum. The phase angle of each transmission carrier is then evaluated and converted back to the data word by demodulating the received phase. The data words are then combined back to the same word size as the original data and converted back to original binary form.

4.4 Calculation of PAPR and CCDF of Original OFDM Signal

The Peak-to-Average Power Ratio (PAPR) is a measurement of a waveform, calculated from the peak amplitude of the waveform divided by the Root Mean Square (RMS) value of the waveform.

The Peak-to-Average Power Ratio (PAPR) of the original Orthogonal Frequency Division Multiplexing (OFDM) signals can be calculated by using the equation (4.4.1).

$$PAPR = \frac{\max_{0 \leq n \leq N-1}|x_n|^2}{E[|x_n|]^2} \qquad (4.4.1)$$

Where, $PAPR$ = Peak-to-Average Power Ratio

x_n = Oversampled OFDM signal

$max\ 0 \leq n \leq N - 1$ = Peak Power

$E[|x_n|]^2$ = Average Power

The oversampled Orthogonal Frequency Division Multiplexing signal, denoted by x_n, is the Inverse Discrete Fourier Transform (IDFT) of the complex data symbols. The oversampled OFDM signal is obtained with the help of the modulation techniques like Quadrature Amplitude Modulation (QAM) or Phase Shift Keying (PSK) at the kth subcarrier.

$$x_n = \frac{1}{\sqrt{N}} \sum_{k=0}^{N-1} X_k e^{j2\pi \frac{k}{N}n}, n = 0, 1, 2, 3, \ldots, N-1 \quad (4.4.2)$$

Where, x_n = Oversampled OFDM Signal

N = Number of Subcarriers

X_k = Complex Data Symbols using PSK or QAM at the kth Subcarrier

The Peak-to-Average Power Ratio (PAPR) of the Orthogonal Frequency Division Multiplexing (OFDM) systems in dB is given by the equation (4.4.3).

$$10\ PAPR\ in\ dB = 10log\ (PAPR) \quad (4.4.3)$$

Where, PAPR = Peak-to-Average Power Ratio

4.5 Complimentary Cumulative Distribution Function (CCDF)

The Peak-to-Average Power Ratio (PAPR) of any OFDM signal is to be calculated in terms of the Complimentary Cumulative Distribution Function (CCDF) as the CCDF is one of the most frequently used performance measures for the PAPR reduction techniques.

For every real number x, the cumulative distribution function (CDF) of a real-valued random variable X is given by

$$F_X(x) = P(X \leq x), \quad (4.5.1)$$

Where the right-hand side represents the probability that the random variable X takes on a value less than or equal to x. The probability that X lies in the interval $[a, b]$, where $a < b$, is therefore,

$$P_r\ (a < X \leq b) = F_x(b) - F_x(a), \quad (4.5.2)$$

Here the notation (a, b), indicates a semi-closed interval.

If treating several random variables $X, Y, ...$ etc. the corresponding letters are used as subscripts while, if treating only one, the subscript is omitted. It is conventional to use F for a cumulative distribution function, in contrast to the lower-case f used for probability density functions and probability mass functions. This applies when discussing general distributions: some specific distributions have their own conventional notation, for example the normal distribution.[20]

The CDF of a continuous random variable X can be defined in terms of its probability density function f as follows:

$$F(x) = \int_{-\infty}^{x} f(t)dt, \qquad (4.5.3)$$

The Complimentary Cumulative Distribution Function (CCDF) describes the probability that a real valued random variable, denoted by X with a given probability distribution will be found at a value greater than x. The Complimentary Cumulative Distribution Function (CCDF) of X is given by the equation (4.5.4).

$$F_c(x) = P(X > x) = 1 - F(x) \qquad (4.5.4)$$

Where, $F_c(x)$= Complimentary Cumulative Distribution Function.

$F(x)$= Cumulative Distribution Function.

In other words, the Complimentary Cumulative Distribution Function (CCDF) of the Peak-to-Average Power Ratio (PAPR) denotes the probability that the PAPR or PAR or Crest Factor of a data block exceeds a given threshold.[20]

4.6 Calculation of SNR and BER of Original OFDM Signal

The Signal-to-Noise Ratio (SNR) and Bit Error Rate (BER) are the other two important parameters to be calculated for the original Orthogonal Frequency Division Multiplexing (OFDM) signal for measuring the performance of the signal. The SNR and BER can be easily calculated by passing the OFDM signal in to the Additive White Gaussian Noise (AWGN) channel, where noise or an unwanted signal gets added to the original OFDM signal.

4.6.1 Additive White Gaussian Noise(AWGN) Channel

Additive White Gaussian Noise (AWGN) channel is a channel model in which the only impairment to communication is a linear addition of wideband or white noise with a constant spectral density (expressed as watts per hertz of bandwidth) and a Gaussian distribution of amplitude. The AWGN channel does not account for fading, frequency selectivity, interference, nonlinearity or dispersion. The wideband Gaussian noise comes from many natural sources such as the thermal vibrations of atoms in conductors, shot noise, black body radiation from the earth and celestial sources such as the Sun [18].

The Additive White Gaussian Noise (AWGN) channel is a good model for many satellite and deep space communication links. The channel is not a good model for most terrestrial links because of multipath, terrain blocking and interference. However, for terrestrial path modeling, AWGN is commonly used to simulate background noise of the channel under study, in addition to multipath, terrain blocking, interference, ground clutter & self interference the modern radio systems encounter terrestrial operation [36].

4.6.2 Signal-to-Noise Ratio (SNR)

Signal-to-Noise Ratio (often abbreviated as SNR or S/N) is a measure used in science and engineering to quantify how much a signal has been corrupted by noise. Signal-to-Noise Ratio (SNR) is defined as the ratio of signal power to the noise power corrupting the signal. A ratio higher than 1:1 indicates more signal than noise.

Signal-to-Noise Ratio (SNR) is sometimes used informally to refer to the ratio of useful information to false or irrelevant data in a conversation or exchange. For example, in online discussion forums, off-topic posts, spam and other online communities are regarded as the noise that interferes with the signal of appropriate discussion.

Signal-to-Noise Ratio (SNR) is defined as the power ratio between a signal i.e., a meaningful information or data and the background noise i.e., an unwanted signal. The SNR is given by the equation 4.6.2.

$$SNR = \frac{P_{signal}}{P_{noise}} (4.6.2)$$

Where, SNR = Signal-to-Noise Ratio

P_{signal} = Signal Power

P_{noise} = Noise Power

In non-technical terms, Signal-to-Noise Ratio (SNR) compares the level of a desired signal (such as music) to the level of background noise. The higher the Signal-to-Noise Ratio, the less obtrusive the background noise is [35].

4.6.3 Bit Error Rate (BER)

The Bit Error Rate or Bit Error Ratio (BER) is defined as the number of bit errors divided by the total number of transferred bits during a studied time interval. BER is a unit less performance measure, often expressed as a percentage number.

In a Communication System, the receiver side BER may be affected by the transmission channel noise, interference, distortion, bit synchronization problems, attenuation, wireless multipath fading etc. The BER may be improved by choosing strong signal strength or by choosing a slow and robust modulation scheme or by line coding scheme or by channel coding schemes.

The transmission Bit Error Rate (BER) is the number of detected bits that are incorrect before error correction, divided by the total number of transferred bits (including redundant error codes). The information Bit Error Rate (BER), approximately equal to the decoding error probability, is the number of decoded bits that remain incorrect after the error correction, divided by the total number of decoded bits (the useful information). Normally, the transmission BER is larger than the information BER. The information BER is affected by the strength of the forward error correction code [33].

4.7 Criteria for selection of PAPR reduction techniques

There are many factors that should be considered before a specific PAPR reduction technique is chosen. These factors include PAPR reduction capability, power increase in transmitted signal, BER increase at the receiver, data rate reduction, computational complexity, and so on. Now we briefly discuss each item[13][20]:

- **PAPR Reduction Capability :**Clearly, this is the most important factor in choosing aPAPR reduction technique. Careful attention must be paidto the fact that some techniques result in other harmfuleffects. For example, the amplitude clipping techniqueclearly removes the time domain signal peaks, but results inin-band distortion and out-of-band radiation.
- **Power Increase in Transmit Signal:**Some techniques require a power increase in the transmit signal after using PAPR reduction techniques. For example, TR technique requires more signal power because some of the transmitted power must be used for the extra subcarrier used for the peak reduction indication. Or TI scheme uses a set of equivalent constellation points for an original constellation to reduce PAPR. Since all the equivalent constellation points require more power than the original constellation,

the transmit signal will have more power after applying TI. When the transmit signal power should be equal to or less than that before using a PAPR reduction technique, the transmit signal should be normalized back to the original power level, resulting in BER performance degradation for these techniques.

- **Loss in Data Rate:** In some techniques the effective data rate is reduced. For example in SLM, PTS, and interleaving techniques, the effective data rate is reduced due to the side information bits used to inform the receiver of what has been done in the transmitter. In these techniques the side information may be received erroneously. Unless some form of protection such as channel coding is employed. In this case the data rate decreases further.

- **Computational Complexity:** Computational complexity is another important consideration in choosing a PAPR reduction technique. For example PTS scheme uses much iteration. The PAPR reduction capability of the interleaving technique is better for a larger number of interleavers. Generally, more complex techniques have better PAPR reduction capability.

- **Low Average power :** Although it also can reduce PAPRthrough average power of the original signals increase, it requires a larger linear operation region in HPA and thus resulting in the degradation of BER performance.

- **No Bandwidth Expansion :** The bandwidth is a rare resource in systems. The bandwidth expansion directly results in the data code rate loss due to side information (such as the phase factors in PTS and complementary bits in CBC). Moreover, when the side information are received in error unless someways of protection such as channel coding employed. Therefore, when channel coding is used, the loss in data rate is increased further due to side information. Therefore, the loss in bandwidth due to side information should be avoided or at least be kept minimal.

- **BER Increase at the Receiver:** This is also an important factor and closely related to the power increase in the transmit signal. Some techniques may have an increase in BER at the receiver if the transmit signal power is fixed or equivalently may require larger transmit signal power to maintain the BER after applying the PAPR reduction technique. For example, the BER after applying ACE will be degraded if the transmit signal power is kept fixed. In some techniques such as SLM, PTS, and selective scrambling, the entire data block may be lost if the side information is received erroneously.

- **No Spectral Spillage :** Any PAPR reduction techniques can not destroy OFDM attractive technical features such as immunity to the multipath fading. Therefore, the spectral spillage should be avoided in the PAPR reduction.
- **Other Factors :** It also should be paid more attention on the effect of the nonlinear devices used in signal processing loop in the transmitter such as DACs, mixers and HPAs since the PAPR reduction mainly avoid nonlinear distortion due to these memories-less devices introducing into the communication channels. At the same time, the cost of these nonlinear devices is also the important factor to design the PAPR reduction scheme.

4.8 Definition of Efficient PAPR

OFDM requires that the whole communication track will be linear within the dynamic range that fits the distribution of PAPR. Usually, as previously discussed there is a problem with Power Amplifier (PA) due to fact that its power efficiency is upper bounded and this upper bound decreases as the PAPR (the dynamic range of the PA) increases. This is the main reason why many PAPR reduction schemes have been proposed [25,26,35] with ultimate goal of reducing the PAPR as much as possible. Such action is not without cost as the Bit Error Rate (BER) of the information bits increases as the reduced PAPR decreases assuming constant Signal to Noise Ratio (SNR).

There are many methods for reducing the PAPR with an ultimate goal of reducing the PAPR as much as possible. If, among other factors, the power efficiency-PAPR relationship of the power amplifier is also taken into account, then there exists a PAPR level for which the BER reaches a minimal value. This PAPR may be used as a definition of an efficient PAPR. The efficient PAPR is not necessary the lowest possible value of PAPR.[19]

CHAPTER 5

PAPR REDUCTION TECHNIQUES

PAPR reduction techniques are classified into the different approaches: Clipping technique, Coding technique, Probabilistic (scrambling) technique, Adaptive Predistortion technique, and DFT-spreading technique

- The clipping technique employs clipping or nonlinear saturation around the peaks to reduce the PAPR. It is simple to implement, but it may cause in-band and out-of-band interferences while destroying the orthogonality among the subcarriers. This particular approach includes block-scaling technique, clipping and filtering technique, peak windowing technique, peak cancellation technique, Fourier projection technique, and decision-aided reconstruction technique
- The coding technique is to select such codewords that minimize or reduce the PAPR. It causes no distortion and creates no out-of-band radiation, but it suffers from bandwidth efficiency as the code rate is reduced. It also suffers from complexity to find the best codes and to store large lookup tables for encoding and decoding, especially for a large number of subcarriers . Golay complementary sequence, Reed Muller code, M-sequence, or Hadamard code can be used in this approach.
- The probabilistic (scrambling) technique is to scramble an input data block of the OFDM symbols and transmit one of them with the minimum PAPR so that the probability of incurring high PAPR can be reduced. While it does not suffer fromthe out-of-band power, the spectral efficiency decreases and the complexity increases as the number of subcarriers increases. Furthermore, it cannot guarantee the PAPRbelowa specified level .This approach includes SLM (Selective Mapping), PTS (Partial Transmit Sequence), TR (Tone Reservation), and TI (Tone Injection) techniques
- The adaptive predistortion technique can compensate the nonlinear effect of a high power amplifier (HPA) in OFDM systems . It can cope with time variations of nonlinear HPA by automatically modifying the input constellation with the least hardware requirement (RAM and memory lookup encoder). The convergence time and MSE of the adaptive predistorter can be reduced by using a broadcasting technique and by designing appropriate training signals
- The DFT-spreading technique is to spread the input signal with DFT, which can be
- Subsequently taken into IFFT. This can reduce the PAPR of OFDM signal to the level of single-carrier transmission. This technique is particularly useful for mobile termi-

nals in uplink transmission. It is known as the Single Carrier-FDMA (SC-FDMA), which is adopted for uplink transmission in the 3GPP LTE standard

5.1 SELECTIVE MAPPING

There are many methods [13, 20] exists for the PAPR reduction. Out of which SLM [20] is a promising technique. According to this technique the main data block will be divided into several independent blocks then each will be converted into OFDM symbol and finally the symbol which has less PAPR will be transmitted. Here,we are showing the performance in baseband transmission.

In this technique, the transmitter generates a set of sufficiently different candidate data blocks, all representing the same information as the original data block, and selects the most favorable for transmission. A block diagram of the SLM technique is shown in Fig. 5.1.1.

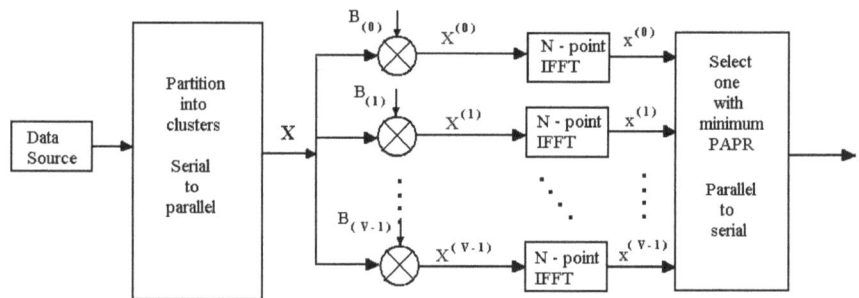

Figure 5.1.1: Block Diagram of Selective Mapping Technique

The classical SLM technique was described in [26]. According to this technique we have to find out independent phase vectors. The number of phase vectors will be same as the number of candidate vectors i.e. let $'M'$.

So the nth point of mth phase vector is given as

$$B(m, n) = e^{\wedge}(j\Phi(m, n)). \qquad (5.1)$$

Where $m \in \{1, 2 \ldots \ldots \ldots M\}$

$n \in \{0, 1 \ldots \ldots \ldots N-1\}$

N = The number of subcarriers

In the classical SLM technique,

$$|B(m, n)| = 1. \qquad (5.2)$$

The random phases $\Phi(m, n)$ will be taken according to [24, 37]. Then we will multiply the data to the phase vectors to find out the candidate vectors and after doing the IDFT operation the OFDM symbol having minimum PAPR is to be transmitted. So to recover the transmitted signal at the receiver, the information about the multiplied phase vector will be required to send along with the OFDM symbol i.e. known as the SI index. If, due to some sort of error the receiver cannot detect the perfect SI index, then the exact transmitted data block from that OFDM will not be recovered.

5.2 Clipping - Based Active Constellation Extension Algorithm

Clipping-Based Active Constellation Extension (CB-ACE) is a basically a combination of two techniques i.e. Active Constellation extension and Clipping. Clipping is simplest technique for PAPR Reduction but a lot of distortion is introduced in it so to make it efficient it needs to be combined with other techniques.

The basic idea was given by Jean Armstrong [23] that is also known as Repeated Clipping and Filtering method (RCF) and to generate the anti-peak signal for reducing the Peak-to-Average Power Ratio (PAPR) by projecting the flipping in-band noise into feasible extension area while removing the out-of-band distortion with filtering [23].

The basic principle of Clipping-Based Active Constellation Extension (CB-ACE) algorithm involves switching between the time domain and the frequency domain. In time domain, clipping is done and then filtering and applying the ACE constraint in the frequency domain because clipping introduces distortion in the signal and both are done repeatedly to supress the to suppress the subsequent regrowth of the peak power [23].

In CB-ACE algorithm the peak amplitude of the original Orthogonal Frequency Division Multiplexing (OFDM) signal is clipping according to a predetermined threshold A and the clipping sample obtained after clipping the peak signals, denoted by $c_n^{(i)}$, is given by the equation below [23].

$$c_n^{(i)} = \begin{cases} \left(\left|X_n^{(i)}\right| - A\right)e^{j\theta_n}, & \left|X_n^{(i)}\right| > A \\ 0, & \left|X_n^{(i)}\right| > A \end{cases}$$

(5.3)

Where, $c_n^{(i)}$ = Clipping Sample of the ith iteration

$X_n^{(i)}$ = Oversampled OFDM signal

A = Predetermined Clipping Level

$\theta_n = \arg(-X_n^{(i)})$

According to the above equation, if the peak amplitude of the original OFDM signal is less than or equal to the predetermined clipping level (A) then the clipping sample is zero and if the peak amplitude of the original OFDM signal is greater than the predetermined clipping level, then the clipping sample is given by $\left(\left|X_n^{(i)}\right| - A\right)e^{j\theta_n}$, where the predetermined clipping level is subtracted from the oversampled OFDM signal an is then multiplied by an exponential value.

The predetermined clipping level, denoted by A, is related to the target clipping ratio, γ and is given by the equation 5.4[23].

$$\gamma = \frac{A^2}{E\{|X_n|^2\}}$$

(5.4)

Where, γ = Target Clipping Ratio

A = Predetermined Clipping Level

X_n = Oversampled OFDM signal

The main drawback of clipping is that it introduces distortion in the signal. This distortion of the original signal can be assumed as the noise, which results to an unreliable communication between the transmitter and the receiver. The distortion caused by clipping the original OFDM signal is categorized into two types, namely [10], [23] –

- In-Band Distortion.
- Out-of-Band Distortion.

The in-band distortion results in the system performance degradation and cannot be reduced, while, the out-of-band distortion can be minimized by filtering the clipped signals. The signal obtained after filtering the clipped signal is given by the equation 5.5.

$$x^{(i+1)} = x^i + \mu \tilde{c}^{(i)}$$

(5.5)

Where, $\tilde{c}^{(i)}$ = Anti-Peak Signal at the ith iteration

μ = Positive Real Number (μ varies from 0.1 to 1)

The anti-peak signal at the ith iteration generated for the PAPR reduction, denoted by $\tilde{c}^{(i)}$, is given by the equation 5.6 [23].

$$\tilde{c}^{(i)} = T^{(i)} c^{(i)}$$

(5.6)

Where, $\tilde{c}^{(i)}$ = Anti-Peak Signal at the ith iteration

$T^{(i)}$ = Transfer Matrix at the ith iteration

$c^{(i)}$ = Peak Signal above the Pre-Determined Level

The transfer matrix at the ith iteration, denoted by $T^{(i)}$, used for generating the anti-peak signal is given by the equation.

$$T^{(i)} = \hat{Q}^{*(i)} \hat{Q}^{(i)}$$

(5.7)

Where, $T^{(i)}$ = Transfer Matrix at the ith iteration

$\hat{Q}^{*(i)}$ = Conjugate of Constellation Order

$\hat{Q}^{(i)}$ = Constellation Order

Though, the process of filtering completely eliminates the distortions caused by the clipping process, it introduces peak regrowth at some of the peak signals of the OFDM signal. The peak regrowth can be reduced by repeating the filtering process, which may again introduce some distortions. Therefore, the clipping and filtering processes are to be repeated until the peak signals are completely reduced. Hence, the Clipping-Based Active Constellation Extension (CB-ACE) Algorithm is also named as the Repeated Clipping and Filtering (RCF) process [23]. All the PAPR Reduction Techniques subsequently results in increase in Bit Error Rate (BER).

5.2.1 Limitations of CB-ACE Algorithm

The Peak-to-Average Power Ratio (PAPR) of the Orthogonal Frequency Division Multiplexing (OFDM) signal is reduced by using Clipping-Based Active Constellation Extension (CB-ACE) Algorithm but has the following limitations or drawbacks [28] –

- Peak regrowth after Digital-to-Analog (D/A) conversion.
- Increase in the Bit Error Rate (BER).
- Out-of-Band Interference (OBI).
- Low clipping ratio problem.

5.3 Exponential Companding Transform

Companding technique is also simple technique in reducing Peak-to-Average Power Ratio (PAPR) which results in signals are uniformly distributed so that the peaks are reduced and thus PAR reduces. There are two types of companding viz Linear and Non-Linear Companding. Linear companding mainly focuses on enlarging small signals but Non-Linear Companding enlarges small signals as well as compresses large signals. This classification is mainly based on the type of companding function used. Non-Linear Companding schemes offer better performance in terms of PAPR Reduction, improved Bit Error Rate (BER) and phase error in OFDM Systems.[21]

The Exponential Companding schemes can effectively reduce PAPR for different modulation formats and sub-carrier sizes. Moreover, many PAPR reduction schemes, such as -law companding scheme, cause spectrum side-lobes generation, but the exponential companding schemes cause less spectrum side-lobes [21].

The Exponential Companding Transform is a type of Nonlinear Companding Transform. The idea of companding comes, from the use of companding in Speech Processing. Since, the Orthogonal Frequency Division Multiplexing (OFDM) signal is similar to that of the speech signal, in the sense that large signals occur very infrequently, the same companding technique can be used to improve the OFDM transmission performance [21].

The key idea of the Exponential Companding Transform is to effectively reduce the Peak-to-Average Power Ratio (PAPR) of the transmitted or the companded Orthogonal Frequency Division Multiplexing (OFDM) signals by transforming the statistics of the amplitudes of these signals into uniform distribution. The uniform distribution of the signals can be obtained by compressing the peak signals and expanding the small signals. The process of companding enlarges the amplitudes of the small signals, while the peaks remain unchanged. Therefore, the average power is increased and thus the Peak-to-Average Power Ratio (PAPR) can be reduced [21].

The Exponential Companding Transform can also eliminate the Out-of-Band Interference (OBI), which is a type of distortion caused by clipping the original OFDM signals. The other advantage of the companding transform is that, it can maintain a constant average power level. The proposed scheme can reduce the PAPR for different modulation formats and sub-carrier sizes without increasing the system complexity and signal bandwidth. The Exponential Companding Transform also causes less spectrum side-lobes [21].

5.3.1 Companding of Original OFDM Signal by using Exponential Companding Transform

The original Orthogonal Frequency Division Multiplexing (OFDM) signal is converted into the companded signal by using the Exponential Companding Transform. The companded signal obtained by using the Exponential or Nonlinear Companding Transform is given by the equation 5.8 [21].

$$h(x) = sgn(x) \sqrt[d]{\alpha[1 - \exp(-\frac{x^2}{\sigma^2})]}$$

(5.8)

Where, $h(x)$ = Companded Signal obtained by Exponential Companding Transform

$sgn(x)$ = Sign Function

α = Average Power of Output Signals

x = Original OFDM Signal

The average power of the output signals, denoted by α, is required in order to maintain the average amplitude of both the input and output signals at the same level. The average power of the output signals is given by the equation 5.9 [38].

$$\alpha = \left(\frac{E[|s_n|^2]}{E\left[\sqrt[d]{1 - \exp\left(-\frac{|s_n|^2}{\sigma^2}\right)}\right]^2} \right)^{\frac{d}{2}}$$

(5.9)

Where, α = Average Power of Output Signals

d = Power of the amplitude of the Companded Signal

5.3.2 Advantages of Exponential Companding Transform

The Exponential Companding Transform reduced the Peak-to-Average Power Ratio (PAPR) of the Orthogonal Frequency Division Multiplexing (OFDM) systems by 7.3 dB at a Complimentary Cumulative Distribution Function (CCDF) of 10^{-2}. The major advantages of the Exponential Companding Transform are [21] –

- PAPR or PAR can be effectively reduced for different modulation formats and subcarrier sizes.
- Average power is maintained at the same level.
- Causes less spectrum side lobes.
- Improved Bit Error Rate.
- Distortion less.

5.3.3 Limitations of Exponential Companding Transform

The Peak-to-Average Power (PAPR) of the Orthogonal Frequency Division Multiplexing (OFDM) systems is reduced by using Exponential Companding Transform but has the only limitation of the transform is the increase in the Bit Error Rate when compared with the original OFDM signal [21].

5.4 Adaptive Active Constellation Extension Algorithm

The main limitation of Clipping-Based Active Constellation Extension(CB-ACE) is that it suffers from low target clipping ratio problem in that they cannot achieve the minimum PAR when the target clipping level is set below an initially unknown optimum value. Moreover, it is difficult to determine the optimal clipping level at the initial stage, because many factors, such as the initial PAR and signal constellation, have an impact on the optimal clipping level determination to solve the low clipping ratio problem of CB-ACE, a technique is introduced which tries to overcome this problem namely Adaptive Active Constellation Extension (Adaptive ACE).[8]

The main objective of the Adaptive Active Constellation Extension (Adaptive ACE) algorithm for reducing the Peak-to-Average Power Ratio (PAPR) is to control both the clipping level and the convergence factor at each step and thereby minimize the peak power signal whichever is greater than the initial target clipping level [8].

The Adaptive Active Constellation Extension (Adaptive ACE) algorithm can be initialized by selecting the parameters namely the target clipping level, denoted by A and the number of iterations, denoted by i. In the first step, the iteration is taken as two i.e., i = 2 and the initial target clipping level is to be taken as A [8].

The predetermined clipping level, denoted by A, is related to the target clipping ratio, γ and given is by the equation 5.10 [8].

$$\gamma = \frac{A^2}{E\{|x_n|^2\}} \quad (5.10)$$

Where, γ = Target Clipping Ratio
A = Predetermined Clipping Level
x_n = Oversampled OFDM signal

The clipping of the peak signal results to distortion of the original OFDM signal. The distortion of the original signal can be assumed as the noise, which results to an unreliable communication between the transmitter and the receiver. The distortion caused by clipping the original OFDM signal is categorized into two types, namely –

I. In-Band Distortion.
II. Out-of-Band Distortion.

The in-band distortion results in the system performance degradation and cannot be reduced, while, the out-of-band distortion can be minimized by filtering the clipped signals. The signal obtained after filtering the clipped signal is given by the equation 5.11 [8].

$$x^{(i+1)} = x^i + \mu \tilde{c}^{(i)}$$

$$(5.11)$$

Where, $\tilde{c}^{(i)}$ = Anti-Peak Signal at the ith iteration
μ = Convergence Factor

The Convergence Factor (CF), denoted by μ can be estimated by using the equation 5.12 [8].

$$\mu = \frac{R[\langle c^{(i)}, \tilde{c}^{(i)} \rangle]}{\langle c^{(i)}, \tilde{c}^{(i)} \rangle} \quad (5.12)$$

Where, μ = Convergence Factor

R = Real Part

$c^{(i)}$ = Peak Signal above the Pre-Determined Level

$\tilde{c}^{(i)}$ = Anti-Peak Signal at the ith iteration

$\langle,\rangle$ = Complex Inner Part

The anti-peak signal at the ith iteration generated for the PAPR reduction, denoted by $\tilde{c}^{(i)}$, is given by the equation 5.13.

$$\tilde{c}^{(i)} = T^{(i)} c^{(i)}$$

(5.13)

Where, $\tilde{c}^{(i)}$ = Anti-Peak Signal at the ith iteration

$T^{(i)}$ = Transfer Matrix at the ith iteration

$c^{(i)}$ = Peak Signal above the Pre-Determined Level

The transfer matrix at the ith iteration, denoted by $T^{(i)}$, used for generating the anti-peak signal is given by the equation 5.14.

$$T^{(i)} = \hat{Q}^{*(i)} \hat{Q}^{(i)}$$

(5.14)

Where, $T^{(i)}$ = Transfer Matrix at the ith iteration

$\hat{Q}^{*(i)}$ = Conjugate of Constellation Order

$\hat{Q}^{(i)}$ = Constellation Order

The original Orthogonal Frequency Division Multiplexing (OFDM) signal, denoted by x_n, is to be clipped in order to reduce the peak signals. The clipping signal is given by the equation 5.15 [28].

$$C_n^{(i)} = \begin{pmatrix} \left|x_n^{(i)}\right| - A \end{pmatrix} e^{j\theta_n}, \left|x_n^{(i)}\right| > A \\ 0, \quad otherwise$$

(5.15)

Where, $C_n^{(i)}$ – Clipping Sample

A – Predetermined Clipping Level

$\theta_n - \arg(-x_n^{(i)})$

The clipping level, denoted by A, for the next iteration is given by the equation 5.16.

$$A^{(i+1)} = A^{(i)} + \mu \nabla_A$$

(5.16)

Where, $A^{(i+1)}$ – Next Iteration Level

$A^{(i)}$ – Present Iteration Level

μ – Convergence Factor

∇_A – Gradient with respect to A

The gradient with respect to the target clipping ratio, denoted by ∇_A, is given by the equation 5.17.

$$\nabla_A = \frac{\sum_{n \in I_1^{(i)} \cup I_3^{(i)}} C_{n3}^{(i+1)}}{N_p}$$

(5.17)

Where, ∇_A – Gradient with respect to A

N_p – Number of peak samples larger than A

The Peak-to-Average Power Ratio (PAPR) is to be calculated to the signal obtained by the equation 4.2, which reduces the PAPR than the PAPR calculated for the original OFDM signal or PAPR obtained of the OFDM signal obtained by using the Clipping-Based Active Constellation Extension (CB-ACE) algorithm [8].

CHAPTER 6

PROPOSED METHOD

6.1 Selected Mapping With Riemann Matrix

Selected mapping (SLM) is a technique used to reduce the peak-to-average power ratio (PAPR) in orthogonal frequency-division multiplexing (OFDM) systems. SLM requires the transmission of several side information bits for each data block, which results in some data rate loss. These bits must generally be channel-encoded because they are particularly critical to the error performance of the system. This increases the system complexity and transmission delay, and decreases the data rate even further. In this work, a novel SLM method for which no side information needs to be sent is proposed. By considering the example of several OFDM systems using either QPSK or 16- QAM modulation, this work shows that the proposed method performs very well both in terms of PAPR reduction and bit error rate at the receiver output provided that the number of subcarriers is large enough[10].

High peak-to-average power ratio (PAPR) is a well-known drawback of orthogonal frequency-division multiplexing (OFDM) systems. As discussed earlier, among all the techniques that have been proposed to reduce the PAPR, selected mapping (SLM) is one of the most promising ones because it is simple to implement, introduces no distortion in the transmitted signal, and can achieve significant PAPR reduction . The idea in SLM consists of converting the original data block into several independent signals, and then transmitting the signal that has the lowest PAPR. The selected signal index, called side information index (SI index), must also be transmitted to allow for the recovery of the data block at the receiver side, which leads to a reduction in data rate. This index is traditionally transmitted as a set of bits (the SI bits). The probability of erroneous SI detection has a significant influence on the error performance of the system since the whole data block is lost every time the receiver does not detect the correct SI index. In practice, a channel code must thus be used to protect the SI bits. This further reduces the data rate, makes the system more complex, and increases the transmission delay. It may therefore be worth trying to implement the SLM method without having to explicitly send any SI bit. A few techniques for doing so have already been proposed such as, for example, the scrambling method described in [10][26].

Generation of phase sequence, which is one of the important aspectsof the SLM technique is very random in existing phase sequence sets.. In the proposed method, the row vectors of the normalised Riemann matrix are selected as the phase sequence set for PAPR reduction in the SLM technique. The Riemann matrix has a definite structure which will be seen in the following Section. Moreover, in all the existing SLM techniques with different phase sequences, reduction in PAPR was less compared to the original OFDM (without any PAPR reduction techniques). In the proposed approach, rows of normalized Riemann matrices (B) are used as phase rotation vectors.

6.2 Concept of Riemann matrices

Riemann matrices arise from the theory of abelian functions. Let us consider $u_1, u_2 \ldots \ldots u_N$ be n independent complex variables and let u=$(u_1, u_2 \ldots \ldots u_N)$. The field of complex numbers is denoted by C, the n-dimensional eulidean space is denoted by C^n. Let $f(u)$ be the abelian function of u; in other words $f(u)$ is a complex-valued function defined and meromorphic in C^n and having 2n periods $\omega_1, \omega_2 \ldots \omega_{2n}$ linearly independent over the field of real numbers. We suppose further that $f(u)$ is a non-degenerate abelian functioni.e.there does not exist any complex linear transformation of the variables $u_1, u_2 \ldots \ldots u_N$ such that $f(u)$ can be brought to depend on strictly less than n complex variables.

    The periods of $f(u)$ form a lattice Γ in C^n, which we may assume, without loss of generality, to be generated by $\omega_1, \omega_2 \ldots \omega_{2n}$ over the ring Z of rational integers. The matrix $P = (\omega_1 \omega_2 \ldots \omega_{2n})$ of n rows and $2n$ columns is called a period-matrix of the lattice Γ. Any other period-matrix P_1 of Γ is of the form $P U$ where U is unimodular.

    The abelian functions admitting all elements of Γ as periods, form a field G. It is known that there exist $n + 1$ abelian functions $f_0(u), f_1(u), \ldots \ldots f_n(u)$ in G such that $f_1(u), \ldots \ldots f_n(u)$ are algebraically indepedent over C, $f_0(u)$ depends algebraically upon $f_0(u), f_1(u), \ldots \ldots f_n(u)$ and further. $G = C(f_0(u), f_1(u), \ldots \ldots f_n(u))$. In other words, G is an algebraic functionfield of n variables over C.

    Let now $\mathcal{L}$ be another field of abelian functions of the form $g(u) = fK^{-1}(u)$ for $f(u) \in G$ and fixed complex nonsingular matrix K. Let us further, suppose that $\mathcal{L}$ has period-lattice Δ contained in Γ. Then it is easy to show that $\mathcal{L}$ is an algebraic extension of G. Moreover, if Q is a period-matrix of Δ, then on the one hand, $Q = KPU$ for a unimodular U and, on the other hand, $Q = PG_1$ for a nonsingular rational integralmatrix G_1. Thus we have

$$KP = PG \quad (6.2.1)$$

with complex nonsingular K and rational integral G. We call any such K a complex multiplication of P and G, a multiplier of P. The object is to study the nature of the setof K and G satisfying the matrix equation (6.2.1). To this end, we first relax our conditions and ask for all rational $2n$ rowed square matrices M satisfying the condition

$$KP = PM \qquad (6.2.2)$$

with a suitable complex matrix K. It is easy to verify that the set of such M is an algebra $\mathcal{M}$ of finite rank over the field Q of rational numbers. We denote this abstract algebra by $\mathcal{M}$ while the set of matrices M give a matrix representation of $\mathcal{M}$ which we denote by $(\mathcal{M})$.

For the period-matrix P, there exists a rational $2n$-rowed alternate non-singular matrix A such that

i. $PA^{-1}P' = 0$ and
ii. $H = \sqrt{-1}PA^{-1}\bar{P}' > 0$ (i.e. positive hermitian) $\qquad (6.2.3)$

We call A, a principal matrix for P.

Any complex matrix P of n rows and $2n$ columns satisfying (6.2.3) for some principal matrix A is called a (n-rowed) Riemann matrix. Alternately, The Riemann matrix (R) is obtained by removing the first row and first column of the matrix A, where [38][39].

$$A(i,j) = \begin{cases} i - 1 & \text{if } i \text{ divides } j \\ -1 & \text{otherwise} \end{cases}$$

If the Riemann matrix (R) is of size $M \times M$, the entries in the normalized Riemann matrix (B) will be $(1/M) R$.

The flow chart of the proposed method is shown in Figure 6.2.1, the algorithm is as follows:

Step 1: Sequences of data bits are mapped to constellation points M-QAM or BPSK to produce sequence symbols $X_0, X_1, X_2, \ldots$

Step 2: These symbol sequences are divided into blocks of length N. N is the number of subcarriers.

Step 3: Each block $X = [X_0, X_1, X_2, \ldots X_{N-1}]$ is multiplied (pointwise multiplication) by U different phase sequence vectors $B^{(u)} = [B_0^{(u)}, B_1^{(u)} \ldots \ldots, B_{N-1}^{(u)}]^T$, where each row of the normalised Riemann matrix B is taken as $B_0^{(u)}, u = 1,2, \ldots \ldots U$

Step 4: A set of U different OFDM data blocks $X^{(u)} = [X_0^{(u)}, X_1^{(u)} \ldots \ldots, X_{N-1}^{(u)}]^T$ are formed, where $X_n^{(u)} = X_n B_n^{(u)}$, $n = 1,2, \ldots \ldots = 1,2, \ldots \ldots U$

Start

Initialize the parameters

Signal Mapping

Divide the symbols into data blocks

Multiply with phase rotation vectors

Tranform into time domain

Select one with lowest PAPR

Stop

Figure 6.2.1 Flow chart of proposed method

Step 5: Transform $X^{(u)}$ into time domain to get $x^{(u)} = IDFT\{X^{(u)}\}$

Step 6: Select the one from $x^{(u)}, u = 1,2,\ldots\ldots U$ which has the minimum PAPR and transmit. The block diagram of the SLM technique is given in figure below:[1]

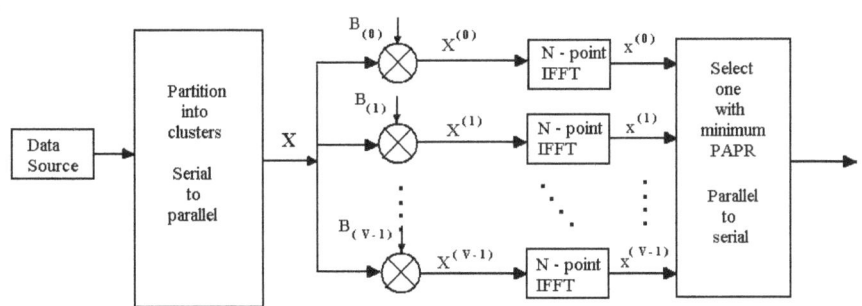

Figure 6.2.2. Block Diagram of proposed SLM Technique

Complexity: In SLM techniques [13, 20, 26], the phase sequences, which are random, have to be sent to the receiver before the actual communication. It is not necessary to send this information to the receiver using the proposed method. This is because the Riemann matrix has a particular structure so the receiver can generate the Riemann matrix. However, compared with other SLM techniques, an extra scaling of $1/M$ for the whole matrix is required for getting the normalized matrix.

CHAPTER 7

RESULTS AND DISCUSSION

7.1 PAPR vs CCDF of Original OFDM Signal

As described earlier that the Peak-to-Average Power Ratio of Original OFDM Signal is found out without imposing any technique. Here in our simulation Matlab is used. Firstly the random data is generated and then after signal mapping to any modulation technique (like BPSK, QPSK and QAM) is followed by IDFT of the generated signal which leads to the generation of an OFDM signal.

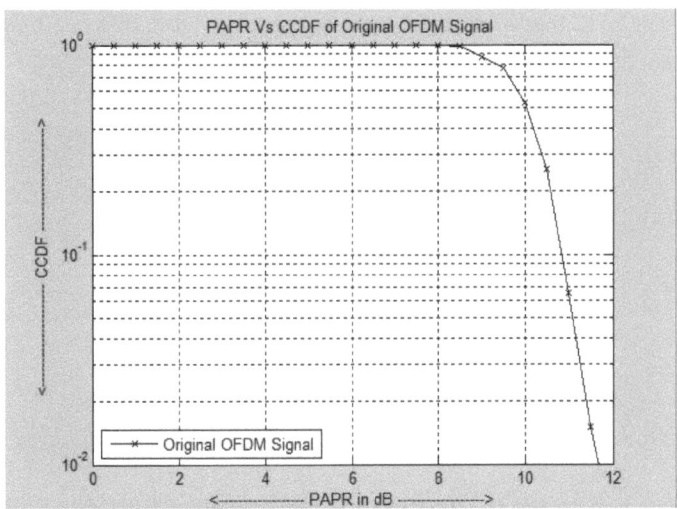

Figure 7.1.1. PAPR Vs CCDF of Original OFDM Signal

From Figure 7.1.1, the Peak-to-Average Power Ratio (PAPR) of the original Orthogonal Frequency Division Multiplexing (OFDM) signal is equal to 11.8 dB with a Complimentary Cumulative Distribution Function (CCDF) of 10^{-2} or 0.01.

The Peak-to-Average Power Ratio (PAPR) of the original Orthogonal Frequency Division Multiplexing (OFDM) signal is very high, which is evident from the figure 7.1.1. The high PAPR results to the increase in the complexity of the Analog-to-Digital Convertors (ADCs) and Digital-to-Analog Convertor (DAC), also reduces the efficiency of the power amplifiers as discussed in previous chapters.

The high Peak-to-Average Power Ratio (PAPR) of the original Orthogonal Frequency Division Multiplexing (OFDM) signals can be reduced by using various reduction techniques like Selective Mapping (SLM), Partial Transmit Sequence (PTS), Interleaving, Tone Reservation (TR), Tone Injection (TI), Active Constellation Extension (ACE), Clipping-Based Active Constellation Extension (CB-ACE) Algorithm, Exponential Companding Transform, Adaptive Active Constellation Extension (Adaptive ACE) Algorithm and New Companding Transform Peak-to-Average Power Ratio (PAPR) of the original Orthogonal Frequency Division Multiplexing (OFDM) signals can be reduced by using various reduction techniques.

7.2 BER of Original OFDM Signal

For finding BER, blocks of OFDM signal are passed and then BER is calculated depending on various SNR's value.

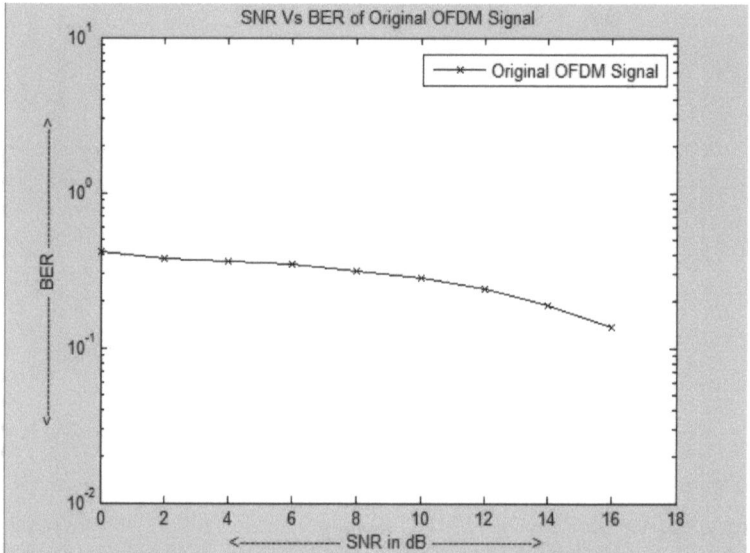

Figure 7.2.1 SNR Vs BER of Original OFDM Signal

From the figure 7.2.1, the Signal-to-Noise Ratio (SNR) of the original Orthogonal Frequency Division Multiplexing (OFDM) signal is equal to 16 dB at a Bit Error Rate (BER) of 10^{-1} or 0.1 approximately i.e., only 1-bit is in error when a stream of 10-bits is transmitted via a communication. Or we can say that, 100-bits are in error when a stream of 1000-bits is transmitted via a communication channel. The Bit Error Rate is less in case of original OFDM Signal than with comparison of calculated when the techniques for PAPR Reduction are used which we will see in later plots.

7.3 PAPR vs CCDF of OFDM Signal by using Selective Mapping (SLM) Technique

Selective Mapping is a technique that converts the original data blocks into various independent signals that have lower PAPR and then selects a block that has lower PAPR is transmitted described in chapter-6.For converting the data block phase rotations are used that are random and then for each and every signal PAPR is calculated and lower PAPR is selected for transmission.

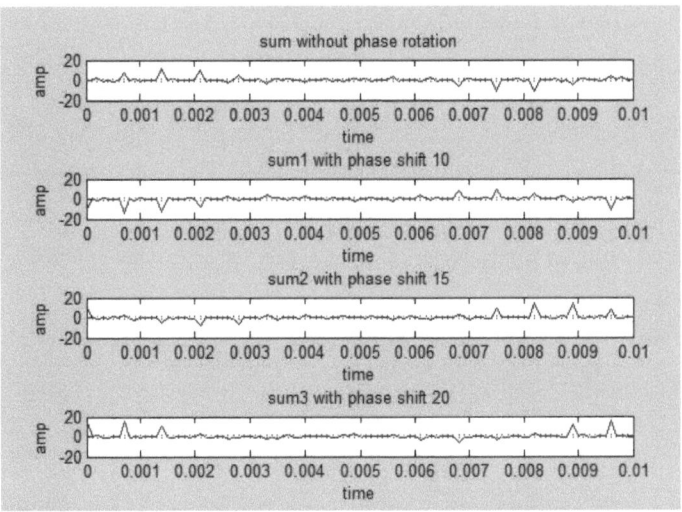

Figure 7.3.1. Independent signals generated with various phase rotations

Figure 7.3.1 shows various independent signals that are generated with various phase rotations and the phase rotations are selected randomly while Figure 7.3.2 shows the calculated PAPR results with these independent signals in Matlab Command Window.

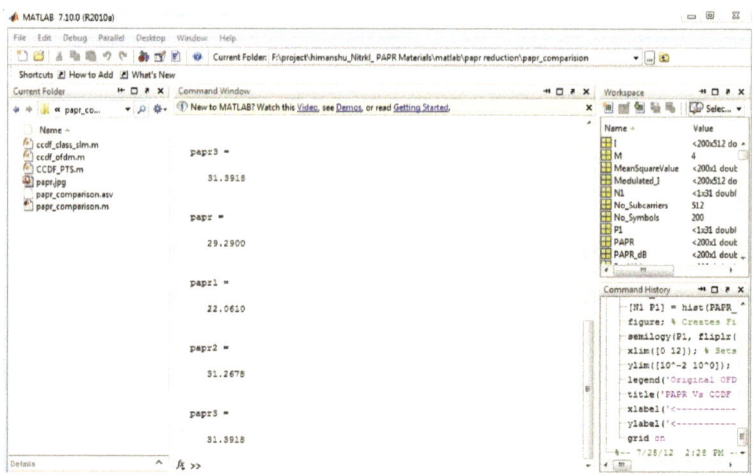

Figure 7.3.2. Calculation of PAPR of independent carriers in SLM technique in Matlab Command Window

Now, the simulation result is shown for various number of subcarriers as shown in figure 7.3.3 .Here, it is seen that for various number of subcarriers the PAPR Reduction capability is different. It is also concluded from the graph that PAPR Reduction is less for large number of subcarriers or in the words, it can be deduced that for large number of subcarriers PAPR is also large.

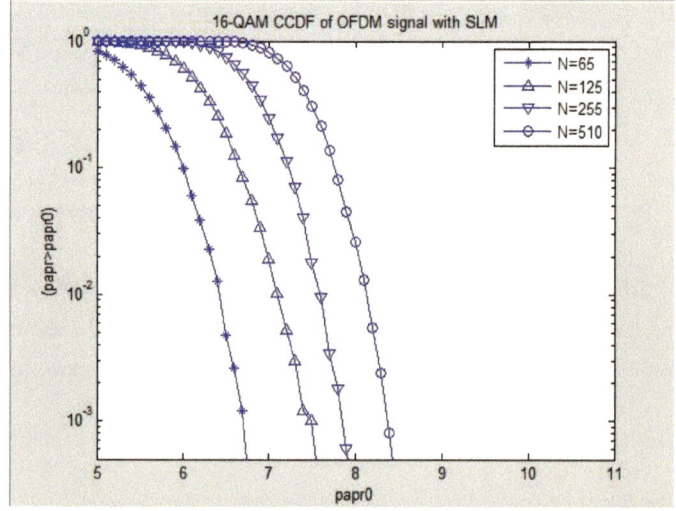

Figure 7.3.3. PAPR vs CCDF plot for SLM Technique for various N

Next, PAPR vs CCDF plot of comparison of original OFDM signal and SLM technique for a fixed number of subcarriers for N= 4 is plotted .It is clear from the graph that for N=4 the PAPR of original OFDM signal is approximately 8.1dB (approx.) and with that of SLM technique the PAPR is found to be 6.6dB (approx.)

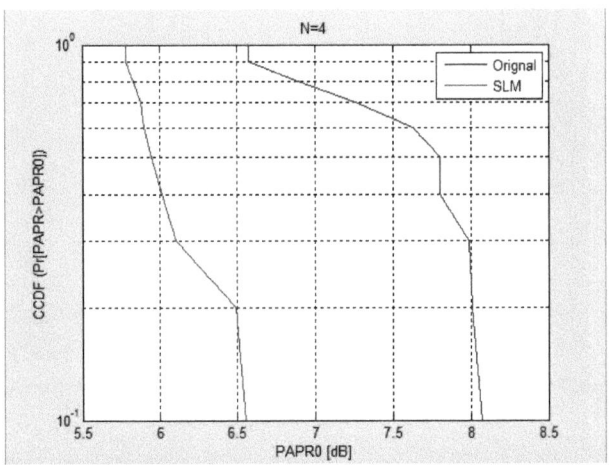

Figure 7.3.4. Comparison of PAPR vs CCDF plot for original OFDM signal and then with SLM technique for N=4.

7.4 CCDF Plot for Clipping-Based Active Constellation Extension (CB-ACE) Technique

Clipping based Active Constellation Extension is based on the combination of two techniques namely Clipping and Active Constellation Extension. Clipping cuts the PAPR signal that is above predetermined threshold level and Active Constellation Extension(ACE) maps the OFDM signals to that of outside signal constellation .In this technique, first clipping and Filtering is done then constraints of ACE is applied .

Figure 7.4.1, it is clear that for CB-ACE technique the PAPR is 10 dB, 8.5 dB and 8.0 dB for the target clipping ratios of 0 dB, 2 dB and 4 dB respectively with a Complimentary Cumulative Distribution Function (CCDF) of 10^{-2} or 0.01.

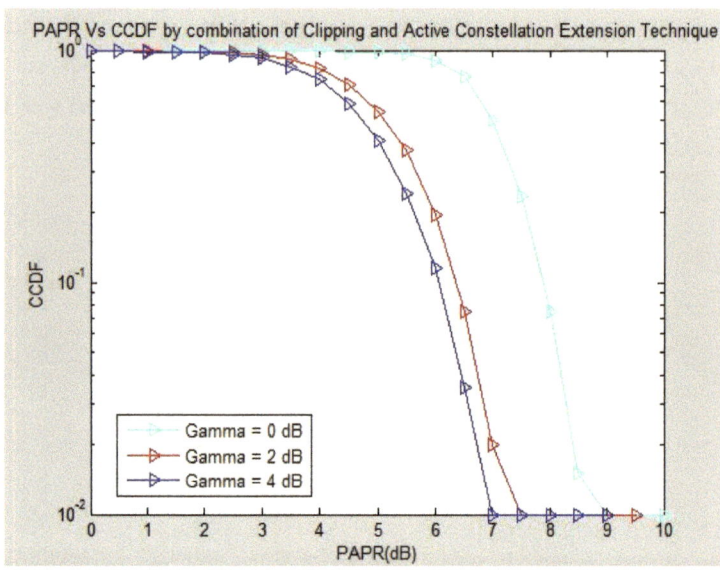

Figure 7.4.1 PAPR vs CCDF plot for combination of Clipping and Active Constellation Extension for various target ratios.

From the same figure, it can be deduced that for CB-ACE technique the PAPR is increasing when the target clipping ratio(Gamma) is decreasing so if we want to decrease PAPR then target clipping ratios should be kept high which is not possible. This issue of CB-ACE technique is known as low target clipping ratio problem which we have tried to overcome with the next technique known as Adaptive Active Constellation Extension (Adaptive- ACE).

There are many other problems also like Out-of-Band Interference (OBI) and peak regrowth. Here, the Out-of-Band Interference (OBI) is a form of noise or an unwanted signal, which is caused when the original Orthogonal Frequency Division Multiplexing (OFDM) signal is clipped for reducing the peak signals which are outside to the predetermined area and the peak regrowth is obtained after filtering the clipped signal. The peak regrowth results to, increase in the computational time and computational complexity.

7.5 BER Plot for Clipping-Based Active Constellation Extension (CB-ACE) Technique

For obtaining BER plot for CB-ACE Technique the signal transmitted via an Additive White Gaussian Noise (AWGN) channel, in order to calculate the Signal-to-Noise Ratio (SNR) and

Bit Error Rate (BER). The BER plot is obtained for different Target Clipping Ratio (Gamma) values as plotted in case of CCDF plot.

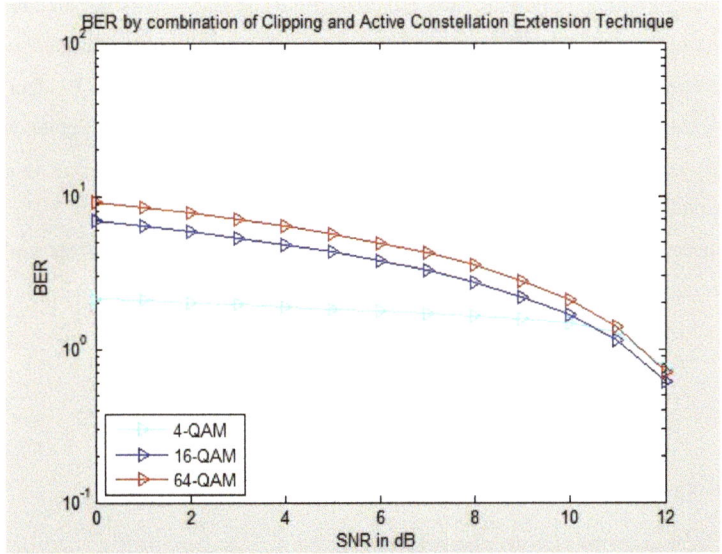

Figure 7.5.1 SNR vs BER plot for combination of Clipping and Active Constellation Extension for various order of modulation techniques.

From the Figure 7.5.1, the Signal-to-Noise Ratio (SNR) of the Orthogonal Frequency Division Multiplexing (OFDM) signal obtained using the Clipping-Based Active Constellation Extension (CB-ACE) algorithm is equal to 12 dB at a Bit Error Rate (BER) of $10^{-0.4}$ for different constellation orders like 4-Quadrature Amplitude Modulation (4-QAM), 16-Quadrature Amplitude Modulation (16-QAM) and 64-Quadrature Amplitude Modulation (16-QAM).

Here, the Bit Error Rate of $10^{-0.4}$ or 0.398, thatmeans a total of 398-bits are in error when 1000-bits are transmitted via a communication channel or approximately 4-bits are in error when 10-bits are transmitted via a communication channel.

7.6 CCDF Plot for Adaptive Active Constellation Extension (Adaptive -ACE) Technique

This technique overcomes the issues by previous technique known as Clipping-Based Active Constellation Extension (CB-ACE). The CB-ACE technique suffers with the problem of low target clipping ratios i.e. PAPR is reduction increases as target clipping ratio increases so he main objective of the Adaptive Active Constellation Extension (Adaptive ACE) algorithm for reducing the Peak-to-Average Power Ratio (PAPR) is to control both the clipping level and the convergence factor at each step and thereby minimize the peak power signal whichever is greater than the initial target clipping level.

For determining optimum threshold level for clipping here the convergence factor is adjusted at each and every iteration which is given by the equation

$$\mu = \frac{R[\langle c^{(i)}, \tilde{c}^{(i)} \rangle]}{\langle c^{(i)}, \tilde{c}^{(i)} \rangle}$$

(7.1)

Where, μ – Convergence Factor

R – Real Part

$c^{(i)}$ – Peak Signal above the Pre-Determined Level

$\tilde{c}^{(i)}$ – Anti-Peak Signal at the ith iteration

$\langle,\rangle$ – Complex Inner Part

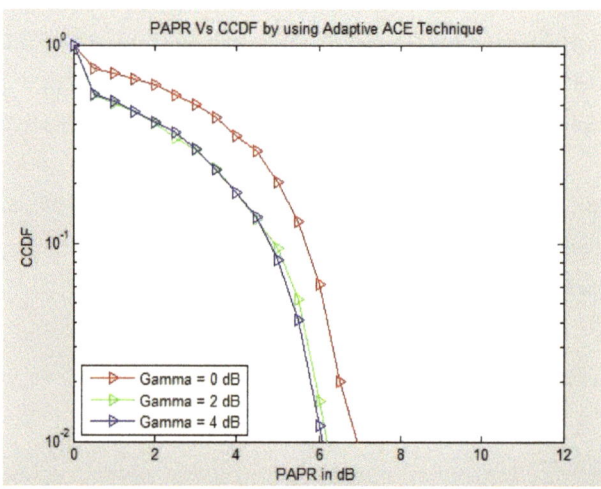

Figure 7.6.1. PAPR vs CCDF plot for Adaptive ACE technique for different target clipping ratios

Therefore for Adaptive-ACE Technique PAPR is equal to 6.8 dB for all the target clipping ratios i.e., for γ = 0 dB or γ = 2 dB or γ = 4 dB with a Complementary Cumulative Distribution Function (CCDF) of 10^{-2} or 0.01 and it is clear from the Figure 7.6.1 that the low target clipping ratio problem of CB-ACE can be overcome by Adaptive-ACE technique.

7.7 BER Plot for Adaptive Active Constellation Extension (Adaptive -ACE) Technique

For obtaining BER plot for Adaptive-ACE Technique the signal is again transmitted via an Additive White Gaussian Noise (AWGN) channel, in order to calculate the Signal-to-Noise Ratio (SNR) and Bit Error Rate (BER). The BER plot is obtained for different Target Clipping Ratio (Gamma) values as plotted in case of CCDF plot.

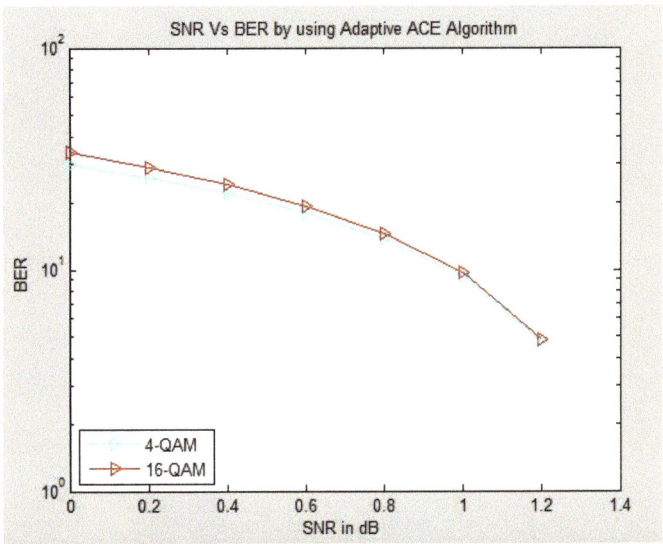

Figure 7.7.1. SNR vs BER plot for combination of Clipping and Active Constellation Extension for various modulation techniques

From the Figure 7.7.1, the Signal-to-Noise Ratio (SNR) of the Orthogonal Frequency Division Multiplexing (OFDM) signal obtained by the Adaptive Active Constellation Extension (Adaptive ACE) algorithm is equal to 1.2 dB at a Bit Error Rate (BER) of $10^{-0.4}$ for different constellation orders like 4-Quadrature Amplitude Modulation (4-QAM) and 16-Quadrature Amplitude Modulation (16-QAM).

Here, the Bit Error Rate of $10^{-0.4}$ or 0.398, thatmeans a total of 398-bits are in error when 1000-bits are transmitted via a communication channel or approximately 4-bits are in error when 10-bits are transmitted via a communication channel.

7.8 CCDF Plot for Exponential Companding Technique

As discussed in chapter-6 the main idea of exponential Companding technique is to make the non-uniform distribution of OFDM signal uniform so that the ratio of peak power to average power is decreased which is same as that of PAPR Reduction. This technique is chosen because it is distortion less and simple so it is suitable for comparison with new techniques. Linear companding schemes focuses just to enlarge small signals but this non-uniform companding scheme focuses on enlarging small signals as well as suppressing large signals also.

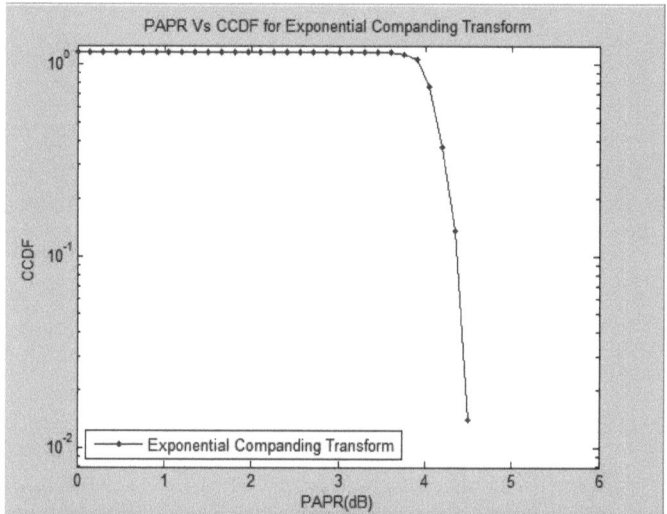

Figure 7.8.1. CCDF plot for Exponential Companding Technique

From the Figure 7.8.1, the Peak-to-Average Power Ratio (PAPR) of the Orthogonal Frequency Division Multiplexing (OFDM) signals obtained by using the Exponential Companding Transform is reduced to 4.5 dB with a Complimentary Cumulative Distribution Function (CCDF) of 10^{-2} or 0.01 which is appreciable with this technique without any distortion. Mow we can see that the PAPR Reduction performance is best in case of this technique than all the techniques described here. The Figure 7.8.1 also shows that the PAPR of OFDM signal is reduced by 7.3 dB at a Complimentary Cumulative Distribution Function (CCDF) of 10^{-2}. This technique is also beneficial as it causes less spectrum side lobes.

7.9 BER Plot for Exponential Companding Technique

From Figure 7.9.1, the Signal-to-Noise Ratio (SNR) of the companded Orthogonal Frequency Division Multiplexing (OFDM) signals obtained by using the Exponential Companding Transform is equal to 12 dB at a Bit Error Rate (BER) of $10^{-0.9}$ for 16-Quadrature Amplitude Modulation (16-QAM).

Here, the Bit Error Rate of $10^{-0.9}$ or 0.126, thatmeans a total of 126-bits are in error when 1000-bits are transmitted via a communication channel or approximately 2-bits are in error when 10-bits are transmitted via a communication channel.

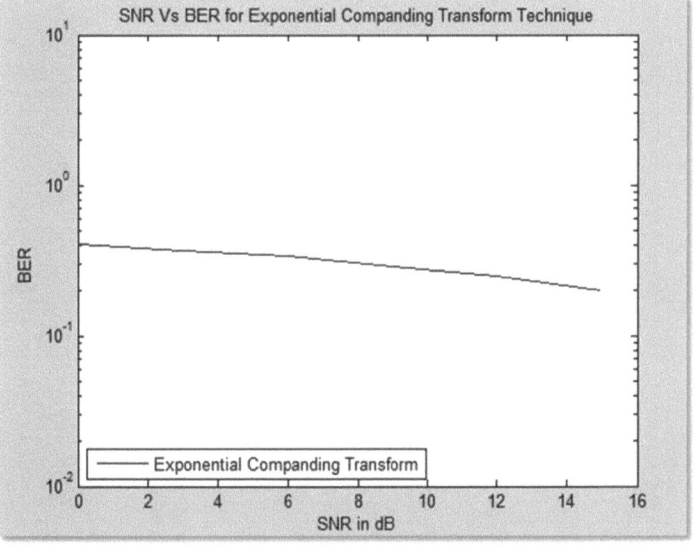

Figure 7.9.1. SNR vs BER by using Exponential Companding Transform

From the above figure it is clear that this technique gives improved BER than all the above techniques mentioned .But The Peak-to-Average Power (PAPR) of the Orthogonal Frequency Division Multiplexing (OFDM) systems is reduced by using Exponential Companding

Transform has the only limitation of the transform is the increase in the Bit Error Rate when compared with the original OFDM signal.

7.10 CCDF Plot for Proposed Technique- SLM with Riemann Matrix

The idea for SLM with Riemann matrix came from a patent of 2004 with Newman phase sequence described in [32][33] where it was proved that with predefined phase sequences called Newman phase sequence, the PAPR was lowest. This proposed technique tries to overcome issues in classical SLM technique .Since in classical SLM technique when the data block are converted randomly into independent signals containing same information and those with lower PAPR is selected for transmission, some side information needs to be transmitted along with the data blocks which increases complexity and lower the data rate as described in chapter-5.In this proposed method, side information is not sent as known data blocks are sent which are rows and columns of Riemann Matrix.

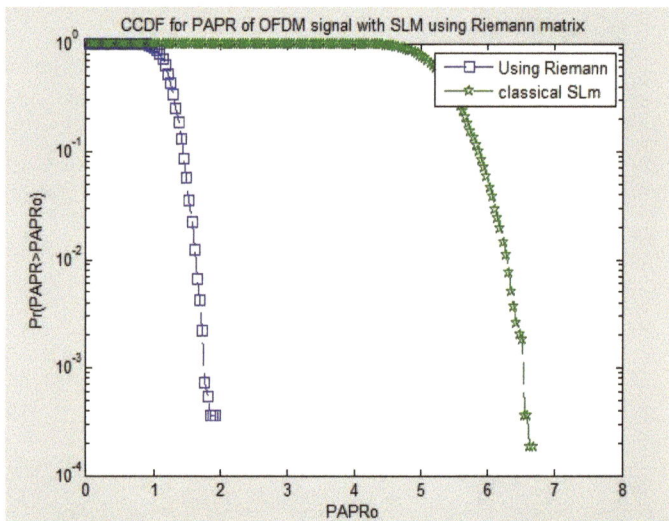

Figure 7.10.1. CCDF plot using SLM with Riemann Matrix

Figure 7.10.1 shows that for classical SLM the PAPR of Orthogonal Frequency Division Multiplexing (OFDM) signal is found to be 6.7(approx.). So, classical SLM technique lowers the PAPR up to 5.1 (approx.). But by using SLM with Riemann Matrix the PAPR of Orthogonal Frequency Division Multiplexing (OFDM) signal is found to be 2.0 (approx.). So, SLM with Riemann Matrix technique lowers the PAPR up to 9.8 (approx.).From the above results it is clear that for SLM with Riemann Matrix technique is better in terms for PAPR Reduction and also it eliminates the major drawback of SLM Technique of sending Side Information (SI Index) Index.

7.11 BER Plot for Proposed Technique- SLM with Riemann Matrix

Figure 7.11.1the Signal-to-Noise Ratio (SNR) of the Orthogonal Frequency Division Multiplexing (OFDM) signals obtained by using the SLM with Riemann Matrixis equal to 12 dB at a Bit Error Rate (BER) of $10^{-0.8}$ for 16-Quadrature Amplitude Modulation (16-QAM).

Here, the Bit Error Rate of $10^{-0.8}$ or 0.158, thatmeans a total of 158-bits are in error when 1000-bits are transmitted via a communication channel or approximately 2-bits are in error when 10-bits are transmitted via a communication channel.

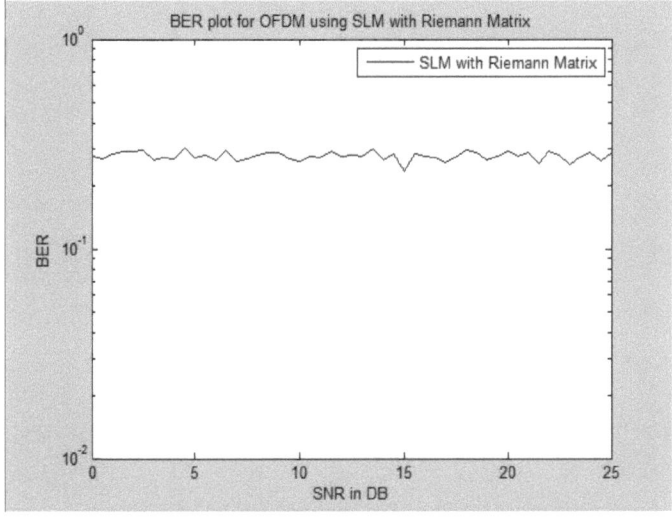

Figure 7.11.1. SNR vs BER by using SLM with Riemann Matrix

From the above figure it is clear that this technique gives improved BER than Clipping-Based Active Constellation Extension (CB-ACE) Technique andAdaptive Active Constellation Extension (Adaptive -ACE) Technique.But The Peak-to-Average Power (PAPR) of the Orthogonal Frequency Division Multiplexing (OFDM) systems is reduced by using Exponential Companding Transform has the only limitation of the transform is the increase in the Bit Error Rate when compared with the original OFDM signal.

CHAPTER 8

CONCLUSION AND FUTURE SCOPE

Orthogonal frequency Division Multiplexing (OFDM) is a very popular and promising technique for the applications of the Wideband Digital Communication Systems such as Digital Audio Broadcasting (DAB), Digital Video Broadcasting (DVB), Wireless Networking, Long Term Evolution (LTE), Broadband Internet Access and Worldwide Interoperability for Microwave Access (WiMAX) etc.

The major drawback of the Orthogonal Frequency Division Multiplexing (OFDM) system is the high Peak-to-Average Power Ratio (PAPR) or Peak-to-Average Ratio (PAR) or Crest Factor. The high PAPR results in the increase in the complexity of Analog-to-Digital Convertors (ADCs) & Digital-to-Analog Convertors (DACs) and also reduces the efficiency of the power amplifiers. Table 1 shows comparison of various PAPR reduction techniques in terms of CCDF

Table 8.1. Comparison of various PAPR reduction techniques in terms of CCDF

Different Techniques	PAPR (in dB)	CCDF
Original OFDM Signal	11.8	10^{-2} or 0.01
Clipping-Based Active Constellation Extension (CB-ACE) Algorithm	10.0 (For $\gamma = 0$ dB) 8.5 (For $\gamma = 2$ dB) 8.0 ((For $\gamma = 4$ dB)	10^{-2} or 0.01
Adaptive Active Constellation Extension (Adaptive ACE) algorithm	6.8 (For $\gamma = 0$ dB, 2 dB or 4 dB)	10^{-2} or 0.01
Exponential Companding Transform	4.5	10^{-2} or 0.01
Selective Mapping	6.9	10^{-4} or 0.0001
Selective Mapping with Riemann Matrix	2(approx.)	**10^{-4} or 0.0001**

The Peak-to-Average Power Ratio (PAPR OFDM signal without any PAPR reduction technique is equal to 11.8 dB (approximately 12 dB) at a Complimentary Cumulative Distribution Function (CCDF) of 10^{-2} or 0.01 while Signal-to-Noise Ratio (SNR) of the original OFDM signal is equal to 16 dB at a Bit Error Rate of 10^{-1} as shown in Table 8.2.

Table 8.2. Comparison of various PAPR reduction techniques in terms of SNR and BER

Different Techniques	SNR (in dB)	BER	No. of Bits in Error (Out of 1000 Bits)
Original OFDM Signal	16	10^{-1}	100
CB-ACE Algorithm	12	$10^{-0.4}$	398
Adaptive ACE Algorithm	12	10-0.4	398
Exponential Companding Transform	12	$10^{-0.9}$	126
Selective Mapping	12	$10^{-0.9}$	126
Selective Mapping with Riemann matrix	**12**	**$10^{-0.8}$**	**158**

The high Peak-to-Average Power Ratio (PAPR) or Peak-to-Average Ratio (PAR) or Crest Factor can be reduced by using various reduction techniques like Clipping-Based Active Constellation Extension (CB-ACE) Algorithm, Exponential Companding Transform, Adaptive Active Constellation Extension (Adaptive ACE) Algorithm and New Companding Transform.

The Clipping-Based Active Constellation Extension (CB-ACE) Algorithm reduces the high Peak-to-Average Power Ratio (PAPR) by clipping and filtering the original OFDM signal. The CB-ACE Algorithm results to peak regrowth, Out-of-Band Interference (OBI), low clipping ratio problem, increase in the Bit Error Rate (BER) and decrease in the Signal-to-Noise Ratio (SNR).

The Active Constellation Extension (ACE) Algorithm provides the minimum Peak-to-Average Power Ratio (PAPR), even when the initial target clipping ratio is set below the unknown optimum clipping point. This algorithm avoids the problem of low clipping ratio, which is caused in the process of reducing the PAPR by using the Clipping-Based Active Constellation Extension (CB-ACE) Algorithm.

The Exponential Companding Transform improves the Bit Error Rate (BER) and minimizes the Out-of-Band Interference (OBI) in the process of reducing the Peak-to-Average Power Ratio (PAPR) effectively by compressing the peak signals and expanding the small signals. The improved BER transmits the data via a transmission channel with fewer errors, while the minimized OBI reduces the effects caused by clipping.

Selective mapping (SLM) is the simplest technique that reduces PAPR without any distortion as with the case of above techniques. The BER is also less as compared to The Active Constellation Extension (ACE) and Clipping-Based Active Constellation Extension (CB-ACE) Algorithm and equal to Exponential Companding Algorithm. The drawback in classical Selective mapping (SLM) technique is sending the side information in the form of Side Information Index (SI Index) which lowers the data rate and also increases complexity. The proposed technique improves classical Selective mapping (SLM) technique and solves the problem of sending side information. Riemann matrix is used in place of Side Information Index (SI Index) which reduces complexity and also maintains the same data rate.

Hence, by reducing the Peak-to-Average Power Ratio (PAPR), the complexity of the Analog-to-Digital Converter (ADC) and Digital-to-Analog Converter (DAC) can be reduced. The reduced Peak-to-Average Power Ratio (PAPR) also increases the efficiency of the Power Amplifiers.

Basic requirement of practical PAPR reduction techniques include the compatibility with the family of existing modulation schemes, high spectral efficiency and low complexity. There are many factors to be considered before a specific PAPR reduction technique is chosen. These factors include PAPR reduction capacity, Power increase in transmit signal, BER increase at the receiver, loss in data rate, computational complexity increase and so on.

Lastly we conclude that no specific PAPR reduction technique is the best solution for all multicarrier transmission systems. Rather, the PAPR reduction technique should be carefully chosen according to various system requirements. In practice, the effect of the transmit filter, D/A converter, and transmit power amplifier must be taken into consideration to choose an appropriate PAPR reduction technique.

By reducing the Peak-to-Average Power Ratio (PAPR), the complexity of the Analog-to-Digital Converter (ADC) and Digital-to-Analog Converter (DAC) can be reduced. The reduced Peak-to-Average Power Ratio (PAPR) also increases the efficiency of the Power Amplifiers.

Recently, multiuser OFDM also has received much attention due to its applicability to high speed wireless multiple access communication systems. In multiuser OFDM system, data streams from multiple users are orthogonally multiplexed onto the downlink and uplink subchannels. In a multiuser OFDM system, a group of carriers is assigned for each user with adaptive modulation, bit and power allocation. Obviously, the characteristics including distribution of the PAPR for each user in uplink multiuser OFDM is the same as that of the PAPR in single user OFDM system since the data of each user will be transmitted to channels independently in uplink multiuser OFDM system. Therefore, the PAPR can be reduced according to these schemes mentioned above in the uplink multiuser OFDM systems.

However, the characteristics of the PAPR in downlink multiuser OFDM is different from that of the PAPR in single user OFDM system since the data composed from different users will be transmitted to channels successively in downlink multiuser OFDM system. Therefore, the PAPR reduction is more complicated in a downlink than that in OFDM uplink in multiuser OFDM systems. If downlink PAPR reduction is achieved by some approaches which have been designed for OFDM, each user has to process the whole data frame and then demodulate the assigned subcarriers to extract their own information. Thus, it introduces additional processing for each user at the receiver. Therefore, based on such type of comparison some modifications in PAPR reduction techniques for the downlink multiuser OFDM systems can be done.

RESEARCH PAPER PUBLISHED

INTERNATIONAL JOURNALS

- "Comparison of Exponential Companding Transform and CB-ACE Algorithm in terms of BER for the PAPR Reduction of OFDM Signal" published in International Journal of Advances in Electronics Engineering (IJAEE), Volume-2, Issue-2,pp .113-116 August, 2012

- "Comparison of Exponential Companding Transform and Adaptive –ACE Algorithm for PAPR Reduction in OFDM signal" published in International Journal of Scientific and Research Publications (IJSRP), Volume 2, Issue 5, May 2012 Edition, ISSN 2250-3153

- "PAPR Reduction of OFDM Based on Adaptive Constellation Extension" published in International Journal of Research in Computer Application and Management (IJRCM),vol no 2 (2012), Issue No. 3(March) ISSN: 2231-1009.

- "Performance of PAPR Reduction Techniques in Terms of BER in LTE-OFDM Signals" published in International Journal of Electronics Communication and Computer Engineering (IJECCE), Volume 3, Issue 3, April 2012 Edition, ISSN 2249-071X

- "Comparison of Exponential Companding Transform and CB-ACE Algorithm for the PAPR Reduction of OFDM Signal" published in GIT-Journal of Engineering and Technology,Volume V, June 2012, ISSN 2249-6157

INTERNATIONAL CONFERENCES

- Mangal Singh, Neelam Dewangan, "Improved Algorithm for the reduction of PAPR in LTE-OFDM Networks" published in the proceedings of 3rd National Conference on Emerging Trends and Application in Computer Science organised by department of Computer Science, St. Anthony's College, Shillong during 30-31st, March, 2012, IEEE Catalog number: CPF1258M-PRT , ISBN : 978-1-4577-0747

- Neelam Dewangan, Suchita Chatterjee and Mangal Singh presented a paper titled "Comparison of Exponential Companding Transform and CB-ACE Algorithm for PAPR Reduction in OFDM" in International Conference on Advances in Computer, Electronics and Electrical Engineering (ICACEEE'12) held in Mumbai on 27th March'2012

- Neelam Dewangan, Mangal Singh, Suchita Chatterjee "PAPR Reduction of SC-FDMA based on Modified Tone Reservation method" in Second International Conference on Control, Communication and Power Engineering, CCPE 2011, organized by the Association of Computer Electronics & Electrical Engineers (ACEEE) - a division of IDES held at Pune on 7-8 November 2011

REFERENCES

[1] Himanshu Bhusan Mishra, Madhusmita Mishra and Sarat Kumar Patra, "Selected Mapping Based PAPR Reduction in WiMAX Without Sending the Side Information" IEEE Conference Publication, March 2012

[2] E. Al-Dalakta, A. Al-Dweik, A. Hazmi, C. Tsimenidis, B. Sharif, "Efficient BER Reduction Technique for Nonlinear OFDM Transmission Using Distortion Prediction" IEEE Transactions on Vehicular Technology, Vol. 61, No. 5, June 2011

[3] Shiann-Shiun Jeng, and Jia-Ming Chen, "Efficient PAPR Reduction in OFDM Systems Based on a Companding Technique with Trapezium Distribution", IEEE Transactions on Broadcasting, Vol. 57, No. 2, June 2011

[4] Yasir Rahmatallah, Nidhal Bouaynaya and Seshadri Mohan, "On The Performance Of Linear And Nonlinear Companding Transforms In OFDM Systems" IEEE Conference on Wireless Telecommunications Symposium (WTS), pp 1-5, April 2011

[5] Suma M N, Kanmani.B, "Developments in Orthogonal Frequency Division Mutiplexing (OFDM) system – A Survey", IEEE Internet (AH-ICI), 2011 Second Asian Himalayas International Conference, pp 1-4, Nov 2011

[6] Jun Hou, Jianhua Ge, Dewei Zhai, Jing Li, "Peak-to-Average Power Ratio Reduction of OFDM Signals With Nonlinear Companding Scheme" IEEE Transactions on Broadcasting, Vol. 56, No. 2, June 2010

[7] Yuan Jiang, "New Companding Transform for PAPR Reduction in OFDM", IEEE Communications Letters, Vol. 14, No. 4, April 2010

[8] Kitaek Bae, Jeffrey G. Andrews and Edward J Powers, "Adaptive Active Constellation Extension Algorithm For Peak-to -Average Power Reduction in OFDM", IEEE Communications letter , Vol. 14 , No. 1 , January 2010.

[9] Ms. V. B. Malode, Dr. B. P. Patil, "PAPR Reduction Using Modified Selective Mapping Technique"Int. J. of Advanced Networking and Applications Vol.02, Issue: 02, pp: 626-630, 2010

[10] Stephane Y. Le Goff, Samer S. Al-Samahi, Boon Kein Khoo, Charalampos C. Tsimenidis, and Bayan S. Sharif, "Selected Mapping without Side Information for PAPR Reduction in OFDM,"IEEE Trans. Wireless Commun., vol. 8, no. 7, pp. 3320-3325, Jul. 2009

[11] Sulaiman A. Aburakhia, Ehab F. Badran, and Darwish A. E. Mohamed, "Linear Com-

panding Transform for the Reduction of Peak-to-Average Power Ratio of OFDM Signals"IEEE Transactions on Broadcasting, Vol. 55, No. 1, March 2009

[12] F.S. Al-kamali, M.I. Dessouky, B.M. Sallam, F. Shawki, F.E. Abd El-Samie, "Transceiver scheme for single-carrier frequency division multiple access implementing the wavelet transform and peak-to-average-power ratio reduction methods" IET Communications, ISSN 1751-8628, 5, 2009

[13] Tao Jiang,Yiyan Wu "An Overview: Peak –to-Average Power Ratio Reduction Techniques for OFDM signals" IEEE Transactions on Broadcasting,Vol.54,No.2,June 2008

[14] Satoshi Kimura, Takashi Nakamura, Masato Saito and Minoru Okada, "PAR Reduction for OFDM signals based on Deep Clipping" ISCCSP 2008, Malta, 12-14 March 2008

[15] Orlandos Grigoriadis, H. Srikanth Kamath, "Ber Calculation Using Matlab Simulation for OFDM Transmission" Proceedings of the International Multi Conference of Engineers and Computer Scientists 2008 ,Vol II, March 2008

[16]Josef Urban ,Roman Marsalek, "OFDM PAPR reduction by combination of Interleaving with Repeated clipping and filtering" IEEE 14th International Workshop on Systems, Signals and Image Processing, 2007 and 6th EURASIP Conference focused on Speech and Image Processing, Multimedia Communications and Services, June 2007

[17] Robert J. Baxley, "Comparing Selected Mapping and Partial Transmit Sequence for PAR Reduction", IEEE Transactions on Broadcasting, Vol. 53, No. 4, December 2007

[18] Hyung G. Myung, Junsung Lim, and David J. Goodman, "Single Carrier FDMA for Uplink Wireless Transmission"; IEEE Vehicular Technology Magazine, pp. 30-38, September 2006.

[19] Dov Wulich, "Definition of Efficient PAPR in OFDM", IEEE Communications Letters, Vol. 9, No. 9, September 2005

[20] Seung hee Han, Jae Hong Lee "An overview of peak-to average power ratio reduction for multicarrier transmission" IEEE Wireless Communications, Vol. 12, Issue 2, pp 56-65 April, 2005.

[21] Tao Jiang ,Yang Yang Yong Hua-Song, "Exponential Companding Technique for PAPR Reduction in OFDM System " IEEE Transactions on Broadcasting ,Vol 51, No. 2, June 2005

[22] J. Armstrong, "Peak-to-Average Power Reduction for OFDM by Repeated Clipping and Frequency Domain Filtering," Elect. Lett., Vol. 38, No. 8, pp. 246–47, Feb. 2002

[23] Jean Armstrong, "New Peak to Average Power Reduction Technique," IEEE Electronic Letters ,Vol .38, No.5 , February 2001

[24] A. D. S. Jayalath, C. Tellambure and H. Wu, "Reduced complexity PTS and new phase sequences for SLM to reduce PAP of an OFDM signal,"IEEE 51st Vehicular Technology Conference Proceedings, 2000. VTC 2000-Spring Tokyo, Vol. 3, pp. 1914-1917, 2000.

[25] Xiaodong Liand, Leonard J. Cimini, "Effects of clipping and filtering on the performance of OFDM," IEEE Communications Letters, Vol. 2, No. 5, MAY 1998.

[26] R. W. Bauml, R. F. H. Fisher, and J. B. Huber, "Reducing the peak-to average power ratio of multicarrier modulation by selected mapping," Electronics Letters, Vol. 32, pp. 2056-2057, Oct. 1996.

[27] Hiroshi Harda, Ramjee Prasad, "OFDM for Wireless Communication System", Arctech House, 2004

[28] M. M. Rana; "Clipping Based PAPR Reduction Method for LTE OFDMA Systems", International Journal of Electrical & Computer Sciences, IJECS-IJENS Vol: 10 No: 05, pp. 1-5.

[29] Stefania Sesia, Issam Toufic, Mattew Baker, "LTE-The UMTS Long Term Evolution", John Wiley and Sons Ltd. ,2009, ISBN 9780470697160

[30] 3GPP Technical Specification Group Radio Access Network, "Requirements for Evolved UTRA (E-UTRA) and Evolved UTRAN (E-UTRAN)," 3GPP, Tech. Rep. TR 25.913, Dec. 2008

[31] Martin Toeltsch, Andreas Molisch, "Efficient OFDM Transmission without Cylic prefix over Frequency-selective channels", IEEE 11th International Symposium on Personal, Indoor and Mobile Radio Communications, September 18-21, 2000, London, (UK).

[32] David Astély, Erik Dahlman, Anders Furuskar, Ylva Jading, Magnus Lindstrom,"LTE: The Evolution of Mobile Broadband"IEEE Communications Magazine April 2009, pp. 44-51

[33] Shashank , Sarat Kumar Patra, "Cross Layer Optimization in OFDM Systems" ,NIT Rourkela, 2012

[34] Yves Louët ,Jacques Palicot, "A classification of methods for efficient power amplification of signals " published by Institute TELECOM and Springer-Verlag France 2008 Ann. Telecommun. (2008) 63:351–368 ,DOI 10.1007/s12243-008-0035-4

[35] H. Ochiai and H. Imai, "Performance analysis of deliberately clipped OFDM signals," IEEE Trans. Commun., vol. 50, pp. 89-101, Jan. 2002

[36] Teddy Purnamirza," The Performance of OFDM In Mobile Radio Channel, University of Technology, Malasia,April 2005

[37] G. Tong Zhou, and Liang Peng, "Optimality Condition for Selected Mapping in OFDM," IEEE Trans. Signal Proces., vol. 54, no. 8, pp. 3159-3164, Aug. 2006.

[38] A.A. Albert, "The principle matrics of Riemann Matrix", Annals of Mathematics ,vol.33, pp 311-318

[39] C. L. Siegel, "Lectures of Riemann matrix", Tata Institute of Fundamental Research, Bombay 1963

[40] Ki-Ho Jung, Heung-Gyooun Ryu, Sung-Ryul Yun, Dong-Kyu Seo, "Apparatus and Method for Transmitting and Receiving Side Information About Selective Mapping in an Orthogonal Frequency Division Multiplexing Communication System" , 2004

[41] Eric Dahlman, Stefan Parkvall , Johan Scold, Per Beming ,"3G Evolution- HSPA and LTE for Mobile Broadband " Second Edition, Academic Press , 2008

[42] Technical white paper on "Long Term Evolution (LTE): A Technical Overview" by Motorola.

[43] Yong Soo Cho, Jaekwon Kim , Won Young Yang , Chung Gu Kang, "MIMO-OFDM WIreless Communications with Matlab" John Wiley and Sons, 2010

[44] B. S. Krongold and D. L. Jones, "PAR reduction in OFDM via active constellation extension," IEEE Trans. Broadcast., vol. 49, no. 3, pp. 258–268, Sep. 2003

[45] Umer Ijaz Butt, "A Study On The Tone-Reservation Technique For Peak-To-Average Power Ratio Reduction In Ofdm Systems", Universal Publication, 2008

[46] Saeed Gazor and Ruhallah AliHemmati, "Tone Reservation for OFDM Systems by Maximizing Signal-to-Distortion Ratio" IEEE Transactions on Wireless Communications, vol. 11, no. 2, February 2012